CONSERVING BIODIVERSITY

A Research Agenda for Development Agencies

Report of a Panel of the Board on Science and
Technology for International Development
U.S. National Research Council

NATIONAL ACADEMY PRESS
Washington, D.C. 1992

NOTICE: The project that is the subject of this report was approved by the Governing Board of the National Research Council, whose members are drawn from the councils of the National Academy of Sciences, the National Academy of Engineering, and the Institute of Medicine. The members of the committee responsible for the report were chosen for their special competence and with regard for appropriate balance.

This report has been reviewed by a group other than the authors according to procedures approved by a Report Review Committee consisting of members of the National Academy of Sciences, the National Academy of Engineering, and the Institute of Medicine.

The National Academy of Sciences is a private, nonprofit, self-perpetuating society of distinguished scholars engaged in scientific and engineering research, dedicated to the furtherance of science and technology and to their use for the general welfare. Upon the authority of the charter granted to it by the Congress in 1863, the Academy has a mandate that requires it to advise the federal government on scientific and technical matters. Dr. Frank Press is president of the National Academy of Sciences.

The National Academy of Engineering was established in 1964, under the charter of the National Academy of Sciences, as a parallel organization of outstanding engineers. It is autonomous in its administration and in the selection of its members, sharing with the National Academy of Sciences the responsibility for advising the federal government. The National Academy of Engineering also sponsors engineering programs aimed at meeting national needs, encourages education and research, and recognizes the superior achievements of engineers. Dr. Robert M. White is president of the National Academy of Engineering.

The Institute of Medicine was established in 1970 by the National Academy of Sciences to secure the services of eminent members of appropriate professions in the examination of policy matters pertaining to the health of the public. The Institute acts under the responsibility given to the National Academy of Sciences by its congressional charter to be an adviser to the federal government and, upon its own initiative, to identify issues of medical care, research, and education. Dr. Stuart Bondurant is acting president of the Institute of Medicine.

The National Research Council was organized by the National Academy of Sciences in 1916 to associate the broad community of science and technology with the Academy's purposes of furthering knowledge and advising the federal government. Functioning in accordance with general policies determined by the Academy, the Council has become the principal operating agency of both the National Academy of Sciences and the National Academy of Engineering in providing services to the government, the public, and the scientific and engineering communities. The Council is administered jointly by both Academies and the Institute of Medicine. Dr. Frank Press and Dr. Robert M. White are chairman and vice chairman, respectively, of the National Research Council.

The Board on Science and Technology for International Development (BOSTID) of the Office of International Affairs, National Research Council, addresses a range of issues arising from the ways in which science and technology in developing countries can stimulate and complement the complex processes of social and economic development. It oversees a broad program of bilateral workshops with scientific organizations in developing countries and conducts special studies.

This report has been prepared by an ad hoc advisory panel of the Board on Science and Technology for International Development. Staff support was funded by the Office of the Science Advisor, Agency for International Development, under Grant No. DAN-5538-G-SS-1023-00.

Library of Congress Catalog Card Number: 91-68333
ISBN: 0-309-04683-1
S-524

Cover Design by David Bennett

Cover art adapted from *Animals: 1419 Copyright-Free Illustrations of Mammals, Birds, Fish, Insects, etc.*, by Jim Harter, Dover Publications, Inc., 1979.

First Printing, January 1992
Second Printing, September 1992

on Biodiversity Research Priorities

PETER RAVEN, Director, Missouri Botanical Garden, *Chairman*
RICHARD NORGAARD, University of California at Berkeley
CHRISTINE PADOCH, New York Botanical Garden, Institute for Economic Botany
THEODORE PANAYOTOU, Harvard Institute for International Development
ALAN RANDALL, Department of Agricultural Economics, Ohio State University
MICHAEL ROBINSON, Director, National Zoological Park, Smithsonian Institution
JAMES RODMAN, Division of Biotic Systems & Resources, National Science Foundation

NRC Staff

JOHN HURLEY, *Director**
MICHAEL MCD. DOW, *Acting Director*
F.R. RUSKIN, *Editor*
JOANN ROSKOSKI, *Senior Project Officer*
CURT MEINE, *Staff Associate and Technical Writer*
NEAL BRANDES, *Program Assistant*
SUSAN PIARULLI, *Program Assistant*

---

* Until November 1991

Preface

In 1980, the National Academy of Sciences published a report entitled *Research Priorities in Tropical Biology*. The authors of that report stated that "destruction of tropical vegetation has attracted the attention of many national and international bodies, especially during the last two decades. If this destruction continues at its present rate until the twenty-first century, it will lead to alteration in the course of evolution worldwide, to widespread human misery, and to loss of the very knowledge that might be used to moderate the other consequences" (NAS, 1980).

Eight years later, the Academy published a book entitled *Biodiversity*, in which its editor, E. O. Wilson, rephrased and broadened the same concern: "Biological diversity must be treated more seriously as a global resource, to be indexed, used, and above all, preserved. Three circumstances conspire to give this matter an unprecedented urgency. First, exploding human populations are degrading the environment at an accelerating rate, especially in tropical countries. Second, science is discovering new uses for biodiversity in ways that can relieve both human suffering and environmental destruction. Third, much of the diversity is being lost through extinction caused by the destruction of natural habitats, again especially in the tropics. We must hurry to acquire the knowledge on which a wise policy of conservation and development can be based for centuries to come" (Wilson, 1988).

Between the publication dates of these two books, literally hundreds of books, articles, and reports were written, and scores of symposia, workshops, meetings, and seminars were organized to publicize the fact that the Earth's biological diversity was decreasing. Locally and internationally financed programs to conserve biodiversity were implemented. The promise of these new efforts, however, is tempered by the reality of the situation: between 1950 and 1990 an estimated 30 to 40 percent of the tropical rain forest disappeared, and with the projected disappearance of an equal amount of forests over the next thirty to fifty years, a quarter of the world's species diversity may vanish forever—more than 2 million species or, on average, 200 per day (Raven, 1988; Wilson, 1989; Ehrlich and Wilson, 1991).

The fate of the tropical forests has come to signify the fate of biodiversity throughout the developing world, indeed throughout the entire world.[1] The successes and failures of conservation efforts over the last decade have demonstrated that the erosion of biological diversity can be stopped only through novel and complex mixtures of economic, social, and political action based on a sound knowledge of ecosystems, including the role of the people who live there. Although we know a great deal about the causes of biodiversity loss, we are only beginning to understand how to formulate development strategies that are based on, and simultaneously conserve, biological diversity. Most people want to protect wildlife and natural resources, but do not know how best to act. At the same time, there is great urgency to act.

This report has been prepared in response to a request from the United States Agency for International Development for a research agenda to provide the critical information needed by decision makers in formulating policy and designing programs to conserve biodiversity. The focus of the study emerged from an organizational meeting, at which it was agreed that the three areas of conservation, economics, and cultural aspects are the most important from the perspective of development agencies. The committee's report draws together findings based on information generated by three meetings of experts in these respective areas.

We stress at the outset that when existing information can be used to implement programs, it should be; the need for more research should not be an excuse to delay action. Although the information and recommendations included here have general relevance for those involved in the conservation of biological diversity, this report is directed specifically toward development agency personnel and development practitioners (nongovernmental and private voluntary organizations).

This audience is important for several reasons. First, biodiversity is greatest in the developing nations of the tropics where development agencies and practitioners expend considerable resources. Through self-examination or persuasion, these agencies are now coming to view conservation as an integral part of the development process (WRI, 1991). The case, however, needs to be strengthened, especially in times of constrained resources and competing demands. All of us must be interested in conservation—locally, regionally, and globally. Because industrial countries use and control a disproportionate share of the world's resources, they must participate in the implementation of sustainable development in the tropics and elsewhere if a new and more enlightened conservation ethic is to take hold.

---

[1] This report is primarily intended for development agencies and therefore concerns developing countries, most of which lie in the intertropical belt. Many developing countries include regions that are not tropical.

Second, industrial countries and development organizations are already devoting substantial resources to these issues, and because the economies in many developing nations are weak, development agencies are likely to be the major funding sources for biodiversity-related activities in the near future. These organizations, however, have limited experience with biodiversity issues and are actively seeking advice on the types of activities they should support.

Third, biodiversity in developing countries cannot be conserved over the long run unless local peoples and national economies simultaneously derive social and economic benefits. This will require not only improved methods of resource management and the creation of products and markets, but the development of economic paradigms involving new institutions, incentives, and policies. International development organizations such as the World Bank and the U.S. Agency for International Development, by the nature of their interaction with the economies of developing nations, will play an important catalytic role in this process.

Decision makers in developing nations constitute an equally important audience for this publication. Biodiversity has been labeled a "developed nation issue." Developing nations often sense that their interests are disregarded by the industrialized world. The developed nations, it has been asserted, are interested in preserving biodiversity only for later exploitation or for global benefit—that is, for their own greater benefit. At the same time, industrial countries have shown a reluctance to change policies and practices that cause environmental degradation and threaten not only their own resource base but the global environment as well. In reality, the biodiversity crisis is a global problem, and each nation or society has a role to play in meeting the challenge ahead. Most developing countries are deeply concerned about the loss of biodiversity but generally lack the means to act on that concern.

In this report, the Panel on Biodiversity Research Priorities has tried to indicate avenues of research that may be helpful in preserving biodiversity and ways in which development agencies can support the people and institutions that study and manage it. This report does not contain a complete action plan or technical research program, nor does it present a detailed discussion of the importance of biodiversity or the reasons for its loss. Numerous recent publications provide comprehensive coverage of these areas (see the References and Recommended Reading at the end of this report). Rather, it presents a research agenda in areas considered of critical importance to the conservation of biodiversity and a discussion of their relevance to sustainable development.

Acknowledgments

The Panel on Biodiversity Research Priorities would like to acknowledge the help of many individuals who participated in planning meetings and reviewed different versions of this report during various stages of its preparation, especially: Peter Ashton, John Briggs, Stephen Brush, Joel Cohen, Herman Daly, John Daly, Douglas Drabkowski, James Duke, James Edwards, Terry Erwin, Rodrigo Gamez, Francesca Grifo, Daniel Janzen, Thomas Lovejoy, Constance McCorkle, Jeffrey McNeely, Kenton Miller, Gene Namkoong, Reed Noss, Bede Okigbo, Christian Orrego, John Pino, Mark Plotkin, Arturo Gómez-Pompa, Kathleen Quick, Terry Rambo, Robert Repetto, Jeffrey Romm, Ulysses Seal, Michael Strauss, Stanley Temple, Bruce Umminger, Patricia Vondal, and D. Michael Warren. Their comments and suggestions were helpful and appreciated. The responsibility for the final report remains wholly ours.

Contents

Executive Summary

The diminishing of the Earth's biological diversity has consequences far more profound than other, sometimes more widely recognized, environmental dilemmas. Because the loss is irreversible—species that are lost are lost forever—the potential impact on the human condition, on the fabric of the Earth's living systems, and on the process of evolution is immense. Our species has evolved biologically and culturally in a highly diverse world. Our interactions with other organisms have shaped our humanity in intricate ways, and our future cannot be separated from that of the life forms and ecosystems with which we share the planet.

The conservation of biodiversity is a global responsibility. Each nation has a necessary role to play in finding new ways to manage biological resources and new ways to sustain commitment. As part of this responsibility, we need to identify what we do *not* know about biodiversity and the means that will be required to increase and disseminate our knowledge.

This report presents an agenda for research in areas critical to the conservation of biodiversity in the world's developing countries. It addresses the biological aspects of conservation as well as the socioeconomic factors and cultural context that must be considered in successful, long-term conservation work in these countries.

The challenge of biodiversity research entails not only the gathering of information, but its management, application, and communication. Likewise, the quality of research depends on the people and institutions that perform it. These considerations are especially important in developing nations and are addressed as part of this agenda. The specific recommendations offered here flow from the general conviction that comprehension and conservation of biodiversity in developing nations represent a challenge of such magnitude that all links in the research chain must be strengthened to ensure success.

The agenda presented here is an ambitious one, but the urgency of the situation requires that it be implemented immediately, and to the fullest extent feasible; a delay of even five years will be too late to prevent irreversible losses. Moreover, we already know a great deal about what needs to be done to preserve biodiversity. We must take immediate steps to reduce losses; we must not wait for research to reveal in full detail how we may sustain biodiversity permanently.

BIOLOGICAL ASPECTS OF CONSERVATION

The state of knowledge of biological diversity suggests that the most basic research requirement is to gain a better, more complete sense of "what's out there." At the same time, we need to know more about how biological diversity is distributed, how it is faring, how to protect it and use it in a sustainable manner, and how to restore it. We also need to improve our ability to gather, organize, communicate, and apply this basic biological knowledge.

Biological Surveys, Inventory, and Monitoring

To achieve an acceptable standard of knowledge about the diversity of the world's biota, the following actions are needed.

National Biological Inventories

National biological inventories should be organized, funded, and strengthened in each country of the world.

This should be the priority for development agencies in biodiversity research. National inventories offer exceptional possibilities for professional linkages and community development and provide the thorough knowledge of organisms necessary for intelligent management of biological diversity to solve any number of practical problems. In many cases this work, with appropriate investments, can be implemented through existing institutions, but should be coordinated through the establishment of national biological institutes or equivalent centers.

Global Biological Survey

A strategy for gauging the magnitude and patterns of distribution of biological diversity on Earth should be coordinated and implemented.

A global survey, drawing on the work of national biological inventories and supplemented by extensive surveys of particular localities, should be undertaken immediately. The National Academy of Sciences study *Research Priorities in Tropical Biology* (NAS, 1980) recommended that a comprehensive, multidisciplinary worldwide survey of well-known groups of tropical organisms (for example, plants, vertebrates, and butterflies) be undertaken. This recommendation is even more timely now. Such a survey would serve as an index to the patterns of species distribution and the nature of communities throughout the tropics, and would provide the cornerstone for a global-scale effort.

Screening of Organisms

The screening of plants, animals, fungi, and microorganisms for features of potential human benefit should be systematized and accelerated through strengthened programs.

Screening allows us to determine more systematically the present and potential uses of organisms for appropriate human purposes. The national biological inventories recommended above should provide both screening opportunities for new natural products and rational methodologies for using materials derived from them.

Monitoring

To detect, measure, and assess changes in the status of biological diversity, appropriate monitoring methods, employing specific indicators of biodiversity attributes, should be implemented.

Inventory, survey, and screening efforts must be complemented by the development and implementation of methods to track the continually changing status of biological diversity. These monitoring efforts should not be undertaken as separate activities, but integrated into the other recommended activities.

Conservation Research

To advance our understanding of successful conservation strategies and methods, the following actions are needed.

Site-Specific Research

To advance the understanding of ecosystem composition, structure, and function; to use this knowledge to link basic and applied research, sustainable land use and development, and the conservation of biological diversity; and to provide baseline data for environmental monitoring, long-term ecological research should be supported at selected sites in developing nations.

Progress toward truly sustainable land use systems requires information on the effect of management options on the ecosystem dynamics, and this information can be gained only through long-term research. Long-term ecological research is especially necessary in the tropical ecosystems of the developing world, where few comprehensive investigations have been undertaken.

Conservation Biology Principles and Methods

Research on biological diversity in developing countries should focus on the application and further development of the methodologies and principles of conservation biology.

The implementation of conservation strategies in developing nations, particularly the establishment of biological reserves and parks, presents an opportunity to test the sustainability of conservation concepts and practices. Most of these originated in developed nations of the Temperate Zones, where human population pressures are much lighter than in the tropics, and where ecosystems are generally less diverse. Testing and comparing conservation methodologies may enable us to elucidate principles that can be more widely applied.

Sustainable Use of Biological Resources

Research should be conducted on strategies for the sustainable use of biological diversity and for returning something of the value of biodiversity to developing countries.

Sustainable use implies that current human needs should be met without degrading the resource base for future generations. Although many strategies for accomplishing this have been advanced, few have undergone scientific scrutiny. Substantive research results are needed to guide policymakers in choosing among them.

Restoring and Utilizing Degraded Lands

Increased support should be given to research on the restoration and utilization of degraded lands and ecosystems in developing countries.

Restoration of degraded lands, although practiced on a small scale for a number of years in some developed nations, is a relatively new area of emphasis in most developing countries. Currently, only a limited theoretical foundation can be applied to site restoration, and there are very few cases in which these theories have been tested. Development agencies must play a larger role in encouraging these efforts and applying restoration techniques more widely.

Information Needs

To enhance the availability and application of scientific information for the purposes of managing and conserving biological diversity, the following actions are needed.

Computer Data Bases and Inventories

Resources should be devoted to the development of computer data bases, inventories, and information networks for the collection and collation of information. Support should be given to the improvement of interinstitutional coordination, system design, and operational administration through the establishment of national biological institutes or equivalent centers.

As conservation faces greater competition for resources, the need for coordination and shared information to prevent duplication of effort becomes paramount. Researchers and administrators involved in conservation efforts must have access to information on the classification, distribution, characteristics, status, and ecological relationships of species. Much of this information, when it exists, is scattered and difficult to obtain. The development of computer data bases and inventories would be a major factor in overcoming this constraint.

Remote Sensing and Geographic Information Systems

Additional research and technical development are needed to advance the utility of remotely sensed data for ecosystem monitoring in developing countries.

The data of remote sensing techniques, coupled with the data management capacity of geographic information systems, offer unprecedented opportunities to assess and monitor ecosystem processes. Even regions that are experiencing rapid change, such as tropical environments, can be closely surveyed through means not available a decade ago. However, remote sensing data must be made more available in developing countries, and training opportunities must be increased.

Strengthening Scientific Networks

Development agencies should use their financial and institutional resources to establish and encourage networks that foster communication among scientists working with biological diversity in developing countries.

As the need for scientific information on biological diversity grows, and as the volume and the quality of information increase, scientific networks must keep pace. These networks should serve to improve communication among scientists in developing countries, between scientists in different countries, and between scientists in both developing and developed countries.

Human Resources

To strengthen the human resources necessary to survey, research, monitor, and manage biological diversity in developing nations, the following actions are needed.

Developing Taxonomic Expertise

International development agencies should sponsor and support the development of taxonomic expertise, both paraprofessional and professional, as an increasingly important part of their conservation programs.

Many of the recommendations previously outlined presume the existence of the taxonomic expertise to implement them. Yet the cadre of trained taxonomists able to perform this work simply does not exist. To describe, inventory, classify, monitor, and manage biological diversity, such expertise must be cultivated.

Strengthening Local Institutions

Because the fate of biological diversity in developing countries depends ultimately on the sense of stewardship, scientific capacities, and administrative structures within these countries, it is important that development agencies invest in strengthening local institutions.

Only native institutions are capable of imparting the understanding of biological diversity among the general public and the proficiency among professionals that will result in effective conservation. It is especially important that development agencies support nongovernmental organizations, educational institutions, museums, and libraries in developing countries, and foster effective operation of the government agencies legally charged with managing resources.

Expanding Cooperative Research Programs

New and existing programs of international cooperative research should undertake research on biological diversity as a fundamental part of their mission, and should be given the financial and administrative support to do so.

New and existing international cooperative research programs should devote more attention to research on biological diversity and should emphasize increased levels of cooperation with developing countries. Biodiversity and its relationship to sustainable land use are central to attaining development goals and should be fundamental considerations in carrying out all research programs involving natural resource management. In particular, these programs need to involve more systematists and other biologists to perform basic research on biodiversity.

SOCIOECONOMIC CONTEXT

The accelerated rate at which the world's biological diversity is being eroded can be attributed, in large part, to socioeconomic factors that encourage exploitative development practices while discouraging conservative resource use. The economic aspects of biodiversity conservation in developing countries demand sophisticated analysis, necessarily involving economists and ecologists working together and with other researchers. The overall objectives of an economic research agenda are: to identify the economic forces leading to the loss of biodiversity within a country; to determine the role of international economic institutions and trends that support this depletion; to elucidate the principles operant in cases of successful development and conservation; and to develop and test economically viable mechanisms for slowing resource depletion and stimulating conservation.

Project- or Country-Level Research

Project- or country-level socioeconomic research is considered urgent. Emphasis should be placed on three critical areas of inquiry: causal mechanisms, valuation, and incentives or disincentives.

- Causal Mechanisms. Economic instruments (including rents, taxes, royalties, concessions, tax holidays, low- or no-interest loans, government-financed infrastructure development, and land tenure systems based on landscape alteration) are integral components of development programs and should be analyzed carefully to determine their short- and long-term effects on natural resource depletion rates.

- Valuation. Because the full value of biodiversity is not yet recognized or incorporated in the policy process, valuation research must be given high priority.

- Incentives and Disincentives. Economic incentives and disincentives play an important role in inducing local people, governments, and international organizations to conserve—or deplete—biological diversity. Research needs to focus on their application and effect.

International Economic Research

Many of the economic forces that profoundly affect the depletion or conservation of biodiversity within a country are transnational in origin. However, the relationships between these transnational forces and natural resource exploitation within countries are poorly understood. Therefore, more research is needed that focuses on the interaction between national and international economic factors, institutions, trends, and impacts.

Global Macroeconomic Research

Because biodiversity is important at the global as well as the local and national levels, many nations and institutions will have to contribute if conservation efforts are to succeed. Development agencies often have central bureaus that attempt to anticipate global needs and to apply a global perspective when designing their in-country activities. These bureaus should support research that focuses on macroeconomic forces operating on a global scale and establishes principles of biodiversity conservation that can be applied in most countries.

CULTURAL CONTEXT

Biological diversity has been lost as a result of social processes, and will ultimately be conserved only through adjustments in these processes. Unless and until they are understood, there is little lasting hope for conservation. In developing countries, the fundamental challenge for researchers in the social sciences is to determine if, where, and how complex local systems can be adapted to modern needs while still retaining the biological diversity of both agroecosystems and surrounding nonagricultural lands. The social sciences can also mediate between indigenous people and institutions, examining indigenous knowledge and land use patterns to understand the problems of biodiversity loss and methods of biodiversity conservation. They can help identify not only the components of local knowledge and land use systems, but also the timing, ecological processes, and structural characteristics that result in the conservation or reduction of biological diversity.

Local Management Systems

Research should provide information on local management systems. We need information on local cultures that are particularly good examples of productive relationships between people and their environment. Special attention should be given to the identification of management systems that are endangered.

Adapting Local Knowledge

Research should promote the application of local knowledge to modern resource management. Based on this information, development agencies would be able to design projects that benefit indigenous people and that benefit from local knowledge. Agencies should identify opportunities to demonstrate how local knowledge can be combined with modern scientific studies in the design of systems for sustainable resource use.

Promoting Local Knowledge

To promote the idea that local knowledge and practices remain relevant for contemporary natural resource management, especially in terms of the scientific insights they provide, the rationale for examining local knowledge and rules should be communicated to professional groups.

Priority Groups

There are, at minimum, thousands of indigenous cultural groups in the developing nations, and it is clearly impossible to study all existing or possible resource use patterns, traditions, combinations, and relationships. A selection of people and places is required. Of highest priority are those use patterns and knowledge systems that are changing most rapidly or disappearing, including those of foragers and collectors, particularly tropical forest dwellers and desert nomadic pastoralists; coastal fisherman, strand foragers, and small island villagers; subsistence agriculturalists raising unconventional staple crops; subsistence agriculturalists raising local cultivars and breeds of conventional crops and animals; and groups that have successfully adapted traditional technologies and resource use patterns in developing market opportunities.

RECOMMENDATIONS

All areas of research on biodiversity and its conservation—biological, economic, and cultural—will require activities across ecological zones and at each level of the "development hierarchy" (project, national, and global). Based on the agenda outlined above, development agencies should place priority on the following actions.

- Development agencies should promote the conservation of biological diversity by strengthening institutional capacity in developing countries. Unless developing countries can build their own corps of competent researchers and develop solid institutions in which they can work, little effective research on biodiversity will be accomplished. The most relevant institutions in this regard are universities, museums, and government ministries, which will require adequate funding, equipment, and personnel to mount national surveys and to establish monitoring capabilities. Information, as it is collected, should be built into data bases so that it can be readily retrieved by other institutions.

- Development agencies should support long-term ecological research sites (comparable to La Selva in Costa Rica or Barro Colorado

Island in Panama) with provisions for continuous monitoring to provide a baseline for understanding natural ecosystems and learning how to modify them most effectively, consonant with development needs.

• Development agencies should place greater emphasis on, and assume a stronger role in, systematizing the local knowledge base—indigenous knowledge, "gray literature," anecdotal information. A vast heritage of knowledge about species, ecosystems, and their use exists, but it does not appear in the world literature, being either insufficiently "scientific" or not "developmental." Much of this information can be interpreted only by local scientists. The U.S. Agency for International Development, the World Bank, and other donor agencies can lend support to the collection and analysis of this important information through their resident offices or missions by providing support to local universities and research institutes as part of project and program development.

• Development agencies must support more and broader-based research in connection with their development projects, including specific studies of the projects' impact on biological diversity, along with the required environmental impact statements. Specific examples of projects that have a direct impact on biodiversity include forest clearing, dam construction, road construction, large-scale human resettlement, and the introduction of new crops or new agricultural production packages and technologies (including the expansion of crop production for export earnings). The impact of these and other activities on ecosystem function and diversity must be investigated more thoroughly, and the likely consequences for affected species must be identified and characterized.

• Development agencies should assist their counterparts in client countries in building the capacity to carry out multidisciplinary research on development options. The capacity to assess and monitor the impact of development activities on biodiversity should receive specific attention. This continuing process should feed its results into the political process at the highest levels—the Ministry of Planning, the Presidency, or equivalent—and provide for the assessment of future economic options in terms of their long-term human and ecological impact. This research should build greater understanding about the relationship between biodiversity and local systems of knowledge and resource use, and should translate this understanding into useful policy and program tools.

• Development agencies should support occasional broader studies of the operation of economic systems as they affect biological diversity. These studies should focus on macroeconomic policy and development strategy—the operation of the economic system locally or regionally—in attempting to provide more generalized conclusions about the relationship between development activities and natural resource management. Studies should analyze the impact of these activities on the conservation of biological resources for agriculture and other economic activities, for their amenity values, and for their influence on future ecosystem stability, including the effects on regional and global climate change, watershed maintenance, river basin flood regimes, and coastal zone (marine, reef, tourism, fishing) resources.

CONCLUSION

In the past decade, the conservation of biological diversity has come to be understood as an essential aspect of sustainable development worldwide. Biodiversity is a basic determinant of the structure and function of all ecosystems and provides the foundation on which the future well-being of human society rests. Research must be expanded and strengthened to improve our understanding of biodiversity, its conservation, and its role in building sustainable human societies.

Many of the nations that are home to the highest concentrations of biological diversity are also crippled by persistent poverty and high rates of population growth, which work against conservation in two ways: (1) by increasing the pressure for inappropriate and harmful land use, and (2) by limiting the ability of individuals and governments to take the steps necessary to halt the degradation of ecosystems and the loss of diversity. International development agencies can and must play a special role in overcoming these obstacles. They are often the most significant sources of funding for human resource development and exert important influence on national and regional policies, economic incentives, and resource use practices that affect the status of biodiversity. Support for research on biodiversity is therefore a critical responsibility of development agencies as they assist client countries in improving the management of their economies and their natural resources.

We need to understand a great deal more about what, why, and how to conserve, and the need is urgent. Although this report focuses on that need, it is also premised on the conviction that research should not serve as a substitute for immediate action to stem the loss of genetic, species, and ecosystem diversity. Rather, research must serve to inform, supplement, and improve these efforts.

1
Biodiversity and Development

We as a species are rapidly altering the world that provides our evolutionary and ecological context. The consequences of these changes are such that they demand our urgent attention. The large-scale problems of unprecedented population growth and inappropriate development are degrading the land, water, and atmosphere, and progressively extinguishing a broad array of the Earth's organisms and the habitats they inhabit. By downplaying these problems or putting them aside in favor of what seem to be more imperative personal, group, or national priorities, we are courting global disaster. By attending to them, we can begin to build a more stable foundation for lasting peace and prosperity.

We live in a world in which far more people are well fed, clothed, and housed than ever before. We also live in a world in which thousands of people, primarily women and young children in developing nations, die each day of starvation or of diseases related to starvation; in which human beings consume well over a third of total terrestrial photosynthetic productivity; and in which human activity threatens, over the next few decades, to eliminate a quarter of the world's species—species we may not use directly, but on which our survival depends in many other ways.

During the 1980s the total human population increased by about 0.8 billion people (from about 4.5 to 5.3 billion), or nearly 2 percent per year. If this rate of growth were to continue, human numbers would double in 39 years (PRB, 1989). If family planning programs and development activities are emphasized consistently and throughout the world, the human population could stabilize, according to United Nations estimates, at about 11 billion by approximately 2090. About 90 percent of this growth is likely to occur in the developing nations. Although population growth may not be the sole cause of environmental degradation, it is almost always an exacerbating factor and undermines the capacity of many developing countries, in particular, to conserve resources and meet basic human needs. As population pressures on land and other natural resources build, the intensity of natural disasters—especially flood and drought—can become aggravated, and the effects more tragic.

There are other, more immediate causes of resource degradation in developing nations, including continuing military conflicts, misguided or misapplied policies that discourage conservation and, above all, persistent and crushing poverty—all of which leave people with few choices in managing land and natural resources. In the past, world leaders in both the developing and the developed nations have tried to address these essentially interrelated problems as separate phenomena. Other global concerns, such as climate change resulting from the buildup of greenhouse gases in the atmosphere, were regarded as separate issues—if they were regarded at all. Few recognized the fundamental need to consider environmental effects and prevent environmental degradation at all stages of development.

Times appear to be changing. The level of concern among world leaders, including the international development agencies, has risen. Many are rethinking their priorities with respect to the allocation of resources to slow the degradation. Whether it is too late for leaders and development agencies to have a beneficial effect depends on what is done and how quickly. Furthermore, this new awareness comes at a time when dramatic political changes in the Soviet Union, Eastern Europe, Central America, the Middle East, and Africa are creating a competing demand for development resources. There are no easy choices, but there can be no turning back to the time when the short-term enrichment of human societies entailed the long-term impoverishment of the living world on which all societies depend.

BIODIVERSITY: DEFINITIONS AND VALUES

The diminishing of the Earth's biological diversity has consequences far more profound than other, sometimes more widely recognized, environmental dilemmas. Because the loss of biodiversity is irreversible—species that are lost are lost forever—the potential impact on the human condition, on the fabric of the Earth's living systems, and on the process of evolution is immense. Our species has evolved biologically and culturally in a highly diverse world. Our past interactions with other life forms have shaped our humanity in intricate ways, and our future cannot be separated from that of the other life forms with which we share the planet.

Biological diversity refers to the variety of life forms, the genetic diversity they contain, and the assemblages they form. Biological systems, whether tundra, forests, savannahs, grasslands, deserts, lakes, rivers, wetlands, coastal communities, or marine ecosystems,

Definitions

Biological diversity (or *biodiversity*, as it has come to be called) refers to the variety and variability among living organisms and the ecological complexes in which they occur. Diversity can be defined as the number of different items and their relative frequency. For biological diversity, these items are organized at many levels, ranging from chemical structures that are the molecular basis of heredity to complete ecosystems. Thus, the term encompasses different genes, species, ecosystems, and their relative abundance (OTA, 1987).

Species is the taxonomic category ranking immediately below genus; it includes closely related, morphologically similar, individual organisms that play a particular ecological role. Species diversity refers to the variety of different species.

Genes represent the basic unit of inheritance, the strands of deoxyribonucleic acid (DNA) polymers that are found in the chromosomes in cell nuclei and control the genetic identity of individual organisms. Genetic diversity refers to the variety of genes.

Species diversity normally refers to the diversity among species, whereas *genetic diversity* refers to the diversity within species.

Ecosystem (derived from "ecological system") refers to the functional system that includes the organisms of a natural community together with their physical environment. Ecosystem diversity is the diversity among systems in a given area.

Evolution is the process of change in the characteristics of organisms by which descendants come to differ from their ancestors.

Biota refers to the collective plant, animal, fungal, and microbial life characterizing a given region.

are functionally complex, and this complexity is associated, in often obscure ways, with the diversity of their component species.

The direct benefits of biological diversity to humanity are myriad. We depend on animal, plant, fungal, and microbial species for food, fuel, fiber, medicines, drugs, and raw materials for a host of manufacturing technologies and purposes. The productivity of agricultural systems is a result of our continual alteration, over thousands of years, of once wild plant and animal germplasm, and still rests on interactions of diverse organisms within agroecosystems. Genetic engineering, especially in the pharmaceutical and food-processing industries, uses natural genetic diversity from sources worldwide. Biomedical research requires comparative information on other species—models such as the mouse and the fruit fly. Although such direct values of biological diversity are not always reflected in market prices, they are more amenable than other values to economic analysis; hence, most economists have focused on this aspect of biological diversity.

Beyond such direct values, biological diversity provides ecological services that are more difficult to calculate with precision. Living organisms are an important part of the processes that regulate the Earth's atmospheric, climatic, hydrologic, and biogeochemical cycles. Only in recent decades have we begun to understand the dynamics of these global processes, and discerning the functional role of biological diversity within them remains a fundamental and challenging question. This is especially important as we seek to understand how biological systems may affect, and be affected by, global climate change resulting from the emission of greenhouse gases into the atmosphere.

It is easier to comprehend (and measure) the ecological services that biological diversity provides more locally in protecting watersheds, cycling nutrients, combating erosion, enriching soil, regulating water flow, trapping sediments, mitigating pollution, and controlling pest populations. As human activities alter landscapes and ecological processes on larger scales, the need for improved management and conservation of land, water, and marine resources will require greater understanding of ecosystem composition, structure, and function. The value of biological diversity in this sense is fundamental.

Finally, ethical and aesthetic concerns direct us to respect, and strive toward rational stewardship of, the world's heritage of biological resources. The noneconomic, intangible, and inherent values of biological diversity take us beyond the traditional realm of the sciences, into the realm of the arts and humanities, language and history, religion and philosophy. These varied modes of human perception and expression have a fundamental stake in the fate of biological diversity, and must contribute to the determination of its fate. Although the values they embody may be less quantifiable, they are nonetheless real and pervasive. To regard biological diversity only for its tangible economic and instrumental values—even where these might be fully taken into account—paradoxically reduces its value.

LOSS OF BIODIVERSITY

The degradation of ecosystems throughout the world, but especially in warmer regions, has been well documented by scientists and is now widely reported in the media. For example, tropical moist lowland forests, which until recently were the least disturbed terrestrial tropical communities, are now experiencing human exploitation on an unprecedented scale. These forests, which may contain more than half of the total species on Earth, have endured longer than other tropical ecosystems because they tend to be difficult to manage. (Deciduous forests, thorn scrub, and other plant communities in the tropics were decimated much earlier.) Their soils are relatively poor in nutrient reserves, often acidic, and subject to rapid leaching of nutrients under the high-rainfall conditions. This makes them relatively difficult to convert to intensive agriculture or forestry systems. Nonetheless, clearing for shifting cultivation, cattle ranching, timber, fuelwood, and conversion to perennial plantations has resulted in the accelerated loss and degradation of primary tropical moist forest. Large areas of the tropics have already been affected. Left unchecked, most of the forests will be entirely lost or reduced to small fragments by early in the next century.

The loss of tropical forest cover can have far-reaching effects, including changes in regional climate (especially rainfall) patterns, changes in biological productivity, accelerated rates of soil erosion, disruption of watershed stability, and increasing emissions of greenhouse gases (which further affects global climate dynamics). In terms of biological diversity, the destruction of primary tropical moist forests causes the extinction of vast numbers of species. Most of the species lost are unknown. Their inherent and aesthetic value, and their potential agricultural, pharmaceutical, or silvicultural values vanish with them.

Although the accelerated pace of deforestation in the humid tropics has drawn widespread attention and is of immediate concern, the degradation of natural ecosystems and habitats, and the loss of their characteristic species diversity, are occurring in nearly every part of the globe as human populations and their support systems expand. We are at a critical juncture for the conservation and study of biological diversity; such an opportunity will not occur again. The Earth's biota is experiencing its greatest episode of species loss since the end of the Cretaceous Era 65 million years ago. More importantly, it is the first mass extinction event that has been caused by a single species—one that we now hope will act, if for no other reason than its own self-interest, to stem the tide (NSB, 1989).

The proximate causes of biodiversity loss are biological, but the root causes of the problem include sociological and economic processes that operate on a global scale. A thorough understanding of the phenomenon will require the investigation and elucidation of both biological and social components, and international cooperation will be necessary to develop both this scientific knowledge and successful mitigation and management strategies. Unless the international community can, indeed, reverse the trend over the next few decades, the erosion of the Earth's biological legacy will continue to accelerate.

Natural Versus Accelerated Rates of Biodiversity Loss

The diversity of life on Earth has never been, and never will be, static. Global biodiversity has fluctuated through geologic time as evolution has added new species and extinction has taken them away. Evolution and extinction are natural processes, the responses of populations of organisms to changes in their physical and biological environment. Change is, in a very real sense, a basic fact of life (Jablonski, 1991).

If change is the norm, why are we now concerned about the conservation of biodiversity? In the past, the environmental changes responsible for fluctuations in diversity occurred over relatively long periods of time. Over the past 15 million years, for example, many parts of the world have gradually become more arid, which has changed the nature of their constituent ecosystems. Even times of relatively rapid environmental change allow organisms the chance to adapt. Over the last 2 million years—a short period by geological standards—glaciers have frequently advanced and retreated, but at a rate gradual enough to allow organisms to migrate and evolve in response. Natural calamities have occasionally destroyed most or even all of one type of ecosystem and great numbers of organisms, but there were always refuges for some species and niches large or small in which evolutionary processes could continue. Even given the role that human beings have had in recent (late Pleistocene and Holocene) extinctions, these have still been isolated, rather than systematic.

The environmental changes affecting biodiversity today have a different origin, order, and magnitude than those recorded in geologic annals. Today, the rate and scale of environmental changes brought about by human activities have increased to the point where a great many species may not have sufficient time or space in which to migrate or adapt.

The current loss of biodiversity has several causes (McNeely et al., 1990; Soulé, 1991). The direct destruction, conversion, or degradation of ecosystems results in the loss of entire assemblages of species. Overexploitation, habitat disturbance, pollution, and the introduction of exotic species accelerate the loss of individual species within communities or ecosystems. More subtly, selective pressures arising directly and indirectly from human activities can result in the loss of genetic variability. Exploitation, habitat alteration, the presence of chemical toxins, or regional climate change may eliminate some genetically distinct parts of a population yet not cause extinction of the entire species. As genetic variability is lost, however, the species as a whole becomes more vulnerable to other factors, more susceptible to problems of inbreeding, and less adaptable to environmental change.

The most important single factor affecting the fate of biodiversity on Earth is the accelerated rate of habitat destruction, particularly in the tropical forests. When an area of forest is cut and the land is converted to intensified use, most of the species living in it cannot survive in the replacement system, be it an agricultural field, pasture, or plantation forest. When any habitat type is reduced to small patches, the organisms that depend on it are in greater danger of extinction as their populations are reduced in number, isolated, and subject to the highly altered impacts of sun, wind, water, soil conditions, other organisms, and human beings. These and other factors enter selectively into small patches of any habitat, severely reducing the diversity of life in that locale (Harris, 1984; Saunders et al., 1991).

In the past, when human activities slowly altered limited areas of the Earth's surface, the rate of local extinctions was barely distinguishable from the natural background rate. Now we may be losing species at a rate 1,000 to 10,000 times greater than the background rate (Wilson, 1988). As Robinson (1988) notes, "We are destroying irreplaceable species on an unprecedented scale without regard for their potential economic, aesthetic, or biological significance." Even conservative estimates of species loss rates suggest that unless current trends are reversed, more than one-quarter of the Earth's species, may vanish in the next 50 years (Raven, 1988; Wilson, 1989; Reid and Miller, 1989; Ehrlich and Wilson, 1991).

Unlike these currently threatened species, or those whose fate is now part of the geologic record, human beings can decide not to choose extinction. We can change our behavior and stop the acceleration of environmental degradation and species loss, thereby safeguarding species, their habitats, and our own future options for their use and enjoyment.

SCIENTIFIC UNDERSTANDING OF BIODIVERSITY

Our understanding of the Earth's biological diversity has significant gaps.* This lack of information hampers our ability to comprehend the

*A recent review of the state of scientific understanding has been provided by the National Science Board (1989) of the National Science Foundation in its report *Loss of Biological Diversity: A Global Crisis Requiring International Solutions*. This report provides the basis for the present discussion (see also Reid and Miller, 1989; Soulé and Kohm, 1989).

magnitude of the loss of biodiversity, prevent further losses, and formulate sustainable alternatives to resource depletion. Answers are still unavailable for seemingly simple but important questions: How many species are there? Where do they occur? What is their ecological role? What is their status—common, rare, endangered, extinct?

Although schemes for classifying organisms date back at least to Aristotle, biologists are still very far from completing an inventory of the Earth's animals, plants, fungi, and microorganisms. The idea of producing encyclopedic treatments of the world's animals and plants began about 300 years ago, toward the close of the seventeenth century. In the eighteenth century, the Swedish naturalist Linnaeus, building on this encyclopedist tradition, devised the system of plant and animal taxonomy involving binomial Latin names that is still used today, in essentially the same form (Mayr, 1982). To date, some 1.4 million kinds of organisms have been assigned scientific names, but coverage is complete for only a few well-studied taxonomic groups such as vertebrates, angiosperms, and butterflies (Wilson, 1988; see table 1-1). Most groups and many major habitats such as coral reefs, the deep sea floor and thermal vents, tropical soils and forest canopies, remain poorly studied. Current estimates of the Earth's total species diversity range from 10 million to 100 million (Wilson, 1988; Ehrlich and Wilson, 1991; Erwin, 1991). Thus, as Wilson (1988) has pointed out, we do not know even to within the nearest order of magnitude the number of species on the planet. Even among those species that have been named, very few have been subject to close biological description or study (NSB, 1989).

Current scientific knowledge, then, is adequate for estimating only the most general characteristics of the abundance and distribution of plants, animals, fungi, and microorganisms of the world. In the following discussions of major taxonomic groups, aquatic systems, and marine biota, emphasis is therefore placed less on numbers than on the relative abundance, ecological importance, and economic and scientific significance of organisms.

Plants

Most estimates suggest that there are about 250,000 species of vascular plants in the world. Approximately two-thirds of these are found in the tropics. The New World tropics are particularly rich in species. For example, at least one-sixth of the Earth's diversity of plant life—45,000 species—can be found in Latin America in Ecuador, Peru, and Colombia, which constitute an area about one-third the size of the contiguous United States. There may be twice as many species in Costa Rica, which is about the size of West Virginia, as have been named for the entire tropics of the world (Latin America, Asia, and

TABLE 1-1 Numbers of Described Species of Living Organisms

Kingdom and Major Subdivision	Common Name	No. of Described Species	Totals
Virus			
	Viruses	1,000 (order of magnitude only)	1,000
Monera			
Bacteria	Bacteria	3,000	
Myxoplasma	Bacteria	60	
Cyanophycota	Blue-green algae	1,700	4,760
Fungi			
Zygomycota	Zygomycete fungi	665	
Ascomycota (including 18,000 (lichen fungi)	Cup fungi	28,650	
Basidiomycota	Basidiomycete fungi	16,000	
Oomycota	Water molds	580	
Chytridiomycota	Chytrids	575	
Acrasiomycota	Cellular slime molds	13	
Myxomycota	Plasmodial slime molds	500	46,983
Algae			
Chlorophyta	Green algae	7,000	
Phaeophyta	Brown algae	1,500	
Rhodophyta	Red algae	4,000	
Chrysophyta	Chrysophyte algae	12,500	
Pyrrophyta	Dinoflagellates	1,100	
Euglenophyta	Euglenoids	800	26,900
Plantae			
Byrophyta	Mosses, liverworts, hornworts	16,600	
Psilophyta	Psilopsids	9	
Lycopodiophyta	Lycophytes	1,275	
Equisetophyta	Horsetails	15	
Filicophyta	Ferns	10,000	
Gymnosperma	Gymnosperms	529	
Dicotolydonae	Dicots	170,000	
Monocotolydonae	Monocots	50,000	248,428
Protozoa			
	Protozoans: Sarcomastigophorans, ciliates, and smaller groups	30,800	30,800
Animalia			
Porifera	Sponges	5,000	
Cnidaria, Ctenophora	Jellyfish, corals, comb jellies	9,000	
Platyhelminthes	Flatworms	12,200	
Nematoda	Nematodes (roundworms)	12,000	
Annelida	Annelids (earthworms and relatives)	12,000	
Mollusca	Mollusks	50,000	
Echinodermata	Echinoderms (starfish and relatives)	6,100	
Arthropoda	Arthropods	751,000	
Insecta	Insects		
Other arthropods		123,161	
Minor invertebrate phyla		9,300	989,761
Chordata			
Tunicata	Tunicates	1,250	
Cephalochordata	Acorn worms	23	

TABLE 1-1 Continued

Kingdom and Major Subdivision	Common Name	No. of Described Species	Totals
Vertebrata	Vertebrates		
Agnatha	Lampreys and other jawless fishes	63	
Chrondrichthyes	Sharks and other cartilaginous fishes	843	
Osteichthyes	Bony fishes	18,150	
Amphibia	Amphibians	4,184	
Reptilia	Reptiles	6,300	
Aves	Birds	9,040	
Mammalia	Mammals	4,000	43,853
TOTAL, all organisms			1,392,485

Source: Wilson, 1988.

Africa combined). Although estimates of the total number of plant species are believed to be relatively accurate compared to other groups, more specific biological knowledge is lacking for most plants.

The ability of plants, along with algae and photosynthetic bacteria, to convert radiant energy into chemical energy through photosynthesis places them at the base of all food chains (with the exception of the recently discovered sulfur-reducing chemosynthetic bacteria associated with some deep sea thermal vents). Because many species depend on specific plants for food and other habitat requirements, the destruction of plant diversity threatens much of the diversity of life in general. One-half of the total species diversity of the Earth may be found in the tropical forests and is, therefore, threatened by their destruction or degradation. If current trends continue, almost all the remaining tropical forests will be severely damaged or reduced to small patches within the next few decades, resulting in the extinction of many as yet unknown plant species (Raven, 1988).

The many and varied human uses of plants—as sources of food, medicines, fibers, waxes, oils, and construction materials; as ornamentals; and as providers of a wide range of environmental services—are too numerous to catalog here. It is important to note, however, that new uses for plants are discovered regularly, and research continues to expand our understanding of their role in ecological processes at all levels. Recent interest in taxol, for example, an anti-cancer agent derived from the bark of the Pacific yew (*Taxus brevifolia*), highlights not only our continued reliance on plant-derived drugs, but our lack of knowledge of the biochemical properties of even the well-inventoried plants of the Temperate Zone.

The developing countries, especially those of the tropics, probably harbor many poorly known or as yet undiscovered plant species with properties of potential benefit to society. About 18,000 species of the legume family, for example, have been described, and the family includes many that are widely used for foods, forage, and oils. It also includes many important tropical timber trees. Most legumes form nodules on their roots that harbor bacteria of the genus *Rhizobium*, which are able to convert atmospheric nitrogen directly into a form in which it can be used for plant growth by both the legumes themselves and other organisms. Both the winged bean (*Psophocarpus tetragonolobus*), a food plant native to Papua New Guinea whose use has spread widely through the moist tropics over the past 15 years, and the "wonder tree" ipil-ipil (*Leucaena leucocephala*), native to Central America but carried by the Spaniards to Hawaii and the Philippines, and now hailed as a solution to problems of soil erosion and firewood shortages, are legumes (NRC, 1975, 1979). Legumes are obviously of great economic importance and have significant potential as genetic raw material for agricultural biotechnology. However, most of those that are now used in agroecosystems were discovered quite by chance. Little is being done to investigate the enormous numbers of legume species that exist in the tropics: 6,000 can be found in Latin America alone; of these, an estimated 2,000 or more are threatened with extinction as the forests of Latin America are degraded and disappear. Unless work on these species is undertaken immediately, most will never have been studied in relation to their utility, nor will they have been incorporated into botanical gardens or seed banks.

Although work in plant taxonomy continues, no coordinated effort to inventory the plants of the world has been initiated, and no general data bank exists from which information about such plants can be retrieved. International networks of botanical gardens, seed banks, and other ex situ strategies for preserving plants are in place in some regions but need to be strengthened. Of special concern in this regard is the accelerated loss of genetic diversity in domesticated crops, their varieties and landraces, and their wild relatives. This diversity of germplasm resources has been largely responsible for the gains made in agricultural productivity in recent decades, but even as that diversity is being called upon to meet new agronomic and environmental needs, it faces growing threats (NRC, 1991b).

The expansion of plant inventories, screening, the dissemination of information, and conservation efforts on a global basis—which can build on efforts at the national level—should be matters of high priority, based on our absolute dependence on plants and our ignorance of the properties of most of them. The estimated 250,000 species of plants are manageable in the sense that the status of their population can be monitored relatively easily, and they can be cultivated and reintroduced into the wild where necessary. Progress in all of these efforts, however, is hindered by a lack of financing and by a dearth of scientists trained for systematic studies in tropical countries. The insufficient number of adequately trained scientists makes the preparation of even simple inventories very difficult.

Fungi

We still have much to learn about even the best-known groups of organisms, and for most, our explorations have hardly begun despite the important role they often play in human affairs. Fungi are a case in point. Some fungi cause crop damage costing billions of dollars annually; others are beneficial, for example, in the production of foods and antibiotics, the maintenance of fertile soils, and the decomposition of biomass. Mycorrhizal fungi, which form symbiotic relationships with plant roots and enhance mineral nutrient uptake by their host plants, are critical links between the soil and plant components of most terrestrial ecosystems, and have been shown to have significant impacts on sustainable crop and forest management, as well as on the success of environmental restoration efforts (Harley and Smith, 1983; Miller, 1985; Amaranthus and Perry, 1987; Cook, 1991). These fungi, however, are insufficiently studied, with most attention being devoted to those relatively few associated with economically important plants.

In general, much work is still to be done on the diversity of the world's fungi. Unlike many taxonomic groups, they may reach their highest levels of diversity outside the tropics, in Temperate Zone forests (especially those of the American Pacific Northwest) (Norse, 1990). In the tropics, there is not a single area for which the fungi are even relatively well known, and it is impossible to prepare regional accounts for any but a very few groups on the basis of collections that are available. Even less explored are the ecological roles that fungi play and their potential to develop into pests or to serve as beneficial agents.

Microorganisms

Through the critical role they play in nutrient movement and cycling, microorganisms constitute "biological bridges" between trophic levels, between abiotic and biotic factors, and between the biogeosphere and the atmosphere. The importance of these linkages extends far beyond the microscopic realm that these organisms inhabit. For example, microflora and microfauna contribute to the maintenance of soil fertility and tilth through their ability to catabolize organic matter, produce organic compounds, and control disease outbreaks. Other microorganisms are important sources of greenhouse gases, although research on this aspect of their ecology is still in its early stages. Many types of microorganisms can cause disease in plants and animals. Although diseases are usually considered in terms of their human economic and medical consequences, microbial and parasitic diseases also play a significant role in population regulation within natural communities. Human-induced changes in ecosystems and the resulting alteration in host species abundances can have unforeseen and undesirable effects on the epidemiology of those diseases.

Humans have derived many benefits from scientific knowledge of microorganisms. Actinomycetes alone have been the source of 3,000 antibiotics since 1950 (Demain and Solomon, 1981). In the future, biotechnology promises to increase the use of microorganisms in solving medical, agricultural, and environmental problems. The foundations of research and development in biotechnology are the fundamental understanding and techniques of molecular biology and genetics, and the diversity of naturally occurring organisms. For biotechnology to realize its potential, more knowledge is required about the microorganisms that are the basis for new technologies.

In the past, little funding has been devoted to work in microbial systematics and ecology. In developing countries, the UNESCO-organized network of Microbiological Resource Centres (MIRCENs) helps link scientists in many countries, and serves as a repository of knowledge and germ plasm for microorganisms. However, the resources available to the MIRCENs are woefully inadequate, and they are able to concentrate their efforts only on well-known organisms such as *Rhizobium* and *Frankia*. In general, little is known about the distribution or diversity of microorganisms, much less about their functional role in ecosystems. What we are learning suggests that they are even more important in supporting healthy ecological systems and biological productivity than previously believed.

Improvement in our scientific understanding of microbial ecology will require increased knowledge of microbial systematics—a daunting challenge. Because research on the biology of microorganisms, especially bacteria, involves so much biochemical experimentation, it is expensive. Furthermore, money alone is not the answer. As in other areas of systematic biology, the human resource base here is thin, and institutional support is meager. Rectifying this situation will require attention to education at all levels and to training, retraining, and employment opportunities in universities, agencies, industry, and other organizations.

Invertebrates

Our knowledge of invertebrate species diversity, like that of microorganisms, is poor for most of the world, especially soil and marine environments, and tropical forests. No more than 10 percent of invertebrate species, and probably a far lower percentage, have actually been described. For some groups such as mites and nematodes, taxonomic work has only begun.

The statistics regarding invertebrates are striking. Approximately two-thirds of the 1.4 million described species are invertebrates (Wilson, 1988). Of these, the vast majority are insects. On a single tree in the Tambopata Reserve in Peru, Wilson (1987) collected 43 species of ants belonging to 26 genera. Collections of arthropods from tropical forest canopies have led scientists to suggest that sharply higher estimates of the total number of species on Earth may be warranted (Erwin, 1982, 1983, 1991; Stork, 1988). The biomass figures are equally commanding. For example, ants alone probably comprise between 5 and 15 percent of the biomass of the entire fauna of most terrestrial ecosystems.

Invertebrates play pervasive, though often unseen, roles in many ecosystem functions, including pollination, decomposition, disease transmission, and regulation of other populations. For example, the interactions of soil mesofauna (e.g., nematodes, collembolans, and mites) and soil microorganisms are crucial in maintaining the plant-soil system. Nematodes both feed on and act as dispersal agents for soil bacteria.

Marine invertebrates play major roles in ecosystem function in the ocean, many of which are analogous to those in terrestrial systems (but there are no pollinators). Marine protozoans, as well as crustaceans (e.g., copepods, euphausids, isopods, amphipods, and larvae of other species), link marine primary producers (phytoplankton) with higher levels of the marine food web, such as fish and marine mammals. Some invertebrates (e.g., squid and octopods) feed on or parasitize marine vertebrates. Invertebrates such as corals and some mollusks can substantially modify the physical structure of the marine environment by building reefs. Marine grazers, such as mollusks and echinoderms, can reduce the structural complexity of the marine environment by removing marine macroalgae and angiosperms. Suspension-feeding mollusks and other invertebrates can control particle concentrations in enclosed bodies of water, affecting water turbidity and the water column concentrations of particle-bound elements and compounds. Marine invertebrates also have both positive and negative impacts on humans. Mollusks, crustaceans, and echinoderms are a major source of food in some areas of the world. Some mollusks and echinoderms are used in biomedical research. Invertebrate growth on hard surfaces, such as ships, piers, and buoys, causes major damage each year and humans spend a great deal of money every year to coat marine surfaces with toxicant-fouling materials. Other species actually burrow into wood and rocks, causing structures made of these materials to fail.

Marine invertebrate parasites and disease organisms are not as common as their freshwater and terrestrial counterparts.

The activities of invertebrates can have major economic impacts on humans. Many crops, for example, depend on insect pollinators, yet they can incur significant damage from other insects. Many of the major human diseases—malaria, schistosomiasis, bubonic plague, encephalitis—are caused by or transmitted through invertebrates. For example, the recent spread of Lyme disease in the United States has been linked to ticks that carry the spirochete agent while spending different parts of their life cycles on white-tailed deer and mice.

Abundant as they are, terrestrial invertebrates are also more prone to extinction than most other groups of organisms. Many species are highly specialized with respect to food, habitat, or other environmental requirements and thus are subject to extinction as a result of even relatively small-scale environmental degradation. This is especially true of tropical forest insects, whose ranges are often quite restricted. The alterations of habitat, on all scales, that are taking place in tropical regions thus result in far greater incidence of invertebrate species loss than would alterations on a similar scale in temperate regions.

Studies of invertebrates do not reflect either their numbers or their importance in ecosystems, which represents a primary constraint of biodiversity research as a whole. Invertebrate systematics, especially in the tropical ecosystems of developing countries, is a neglected area in a neglected branch of basic biology. Important taxonomic groups of great diversity are often the responsibility of a handful of resident scientists in tropical countries, while very limited help is available from the large museums of temperate regions. Moreover, many of the present experts are senior scientists whose administrative responsibilities leave them little time for basic taxonomic work (NAS, 1980). Until scientists from temperate and tropical zones alike are encouraged and rewarded for taking up these fundamental taxonomic studies, the lack of trained systematists will be an important limiting factor in the advancement of knowledge on biological diversity.

Vertebrates

As a group, vertebrates have been more thoroughly studied than most other organisms. Approximately 41,000 species have been described, but many have yet to be discovered.

Almost half of the known vertebrates are fish, and most of those that remain undescribed are likely to be fish, primarily because of their relatively inaccessible habitats. For example, it has been estimated that as many as 40 percent of the freshwater fish of South America have not yet been classified scientifically (Böhlke et al., 1978), and the fish of tropical Asia are also poorly known. Data on life-history patterns, food webs, and the behavior of fish are for the most part lacking. Major stocks of many commercial species may be depleted to such an extent in the near future that it will be impossible to study the variety of their adaptations and the conditions under which they evolved. This is especially true with respect to migrating fish that depend on unimpeded access to upper regions of rivers, which are often favored sites for dams. Information of this type is of fundamental ecological and economic importance. Fish also represent a critical human food resource that is insufficiently understood to be used on a fully sustainable basis. In the same sense that tropical rain forests might contain many species whose products could be of great use, fish communities may include members whose nutritional modes, defense mechanisms, behavior, or growth characteristics could be applied in the production of proteins, medicines, or fertilizers, and in the management of aquatic habitats.

In comparison to other taxonomic groups, there are few undescribed species of reptiles, birds, and mammals. Nonetheless, new species continue to be discovered fairly regularly. Even among primates—the most widely and carefully studied group of organisms—new discoveries are still being made. The black-faced lion tamarin, *Leontopithecus caissara*, a previously unknown primate, was discovered in 1990 by two Brazilian biologists on an island close to São Paulo (Lorini and Persson, 1990). Despite our relatively complete knowledge of the species diversity within these groups, we know nothing more about the vast majority of them except that they exist.

Vertebrates—especially those that have been domesticated—are the species of greatest economic and aesthetic importance to human beings. Because much basic zoological research has focused on domesticated vertebrate species and because much of our previous conservation research has focused on wild vertebrate species, these are important models as biodiversity research expands. Moreover, because the highest trophic levels within ecosystems are generally occupied by reptiles, birds, and mammals, efforts to preserve diversity among these groups will have beneficial impacts on other organisms that share—and constitute—their habitat.

Tropical Aquatic Systems

Tropical rivers, lakes, and wetlands are among the richest, most important, yet least studied, habitats in the developing world. The 1980 National Academy of Sciences report *Research Priorities in Tropical Biology* noted the critical scientific and economic importance of these systems, and recommended that they be studied much more intensively and monitored for long-term changes (NAS, 1980). The need for scientific study of these systems, particularly of their biological diversity, has increased in the interim.

Watershed development projects of all kinds inevitably alter river systems and their biota, usually before scientific investigations of unmodified watersheds and basins take place. Research must focus on river systems prior to development if any accurate characterization is to be made of their biological diversity, ecosystem functions, and hydrological dynamics. This need, it should be noted, pertains to rivers in both tropical and nontropical developing countries. In the tropics, it includes both the great rivers—the Amazon, Orinoco, Parana, Zaire, Niger, Nile, Mekong—and the many minor rivers and tributaries.

The needs and opportunities for research in this area are great. The composition, abundance, and functioning of the plankton of large rivers in their natural state are essentially unstudied, and although the opportunity has largely been lost in temperate regions, it is still possible in the tropics (NAS, 1980). The invertebrates of tropical rivers, immense in their variety, are largely unstudied because of the shortage of trained taxonomic experts. Knowledge of the fish and other vertebrates of tropical rivers is somewhat more advanced, but as more systems are altered, the opportunity for comprehensive studies of riverine community structures, trophic interactions, and vertebrate population dynamics becomes increasingly scarce.

Lakes are less common in the tropics than in the temperate zones, primarily because glaciation was a less significant factor in the geological history of the tropics. Nonetheless, the special physical features, high productivity, economic importance, and vulnerability of tropical lakes make the study of their biological diversity particularly important. A number of tropical lakes, large and small, support high levels of fish endemism and merit study not only because of their inherent importance for science, but also because of their susceptibility to the effects of exotic fish introductions. The unique circumstances under which the biota of tropical lakes has evolved and the likelihood of alteration due to development pressures make these lakes important sites for expanded scientific attention. Especially important are Lake Malawi in Africa's Great Rift Valley, Lake Titicaca and smaller lakes of the high Andes, Lake Maracaibo in Venezuela, Lake Toba in Sumatra, and many smaller lakes of insular Southeast Asia (NAS, 1980).

Tropical wetlands, of many varieties, are among the most productive freshwater systems in the world. They are also highly vulnerable to destruction by drainage, conversion to intensive rice production, and the alteration of associated river systems (NAS, 1980). Many of the most important—the Sudd in the Sudan, the Okavango of Botswana, the Pantanal of Brazil, the wetlands of the Sepik and Fly Rivers of Papua

New Guinea—exhibit distinctive species compositions, evolutionary adaptations, energy-flow characteristics, and population dynamics as a result of seasonal fluctuations in water levels and unique chemical factors. Studies of the biological diversity of these systems are critical in understanding how they function, and how human alteration and use of tropical wetlands may affect their diversity and productivity.

Marine Biota

Until recently, interest in biological diversity and its conservation focused primarily on terrestrial and freshwater environments, and thus neglected the most extensive habitat on Earth (Ray, 1988). The very vastness of the marine environment (oceans cover 70 percent of the Earth's surface), the variety of ecosystems it contains, and the difficulties involved in exploring and studying the life of the sea have hampered efforts to treat marine biodiversity more comprehensively. Marine organisms have long been used in cell biology and other areas of basic biological research, and certain communities—in particular, coastal wetlands, mangrove forests, and coral reefs (the species richness of which is often compared to that of tropical rain forests)—have been studied in detail. In general, however, relatively little is known about the diversity, abundance, and distribution of marine organisms or the structure and function of marine ecosystems.

Marine systems are distinguished by their high degree of diversity at all taxonomic levels. Current estimates of the total number of species on the planet assume that approximately 94 percent of the species are terrestrial. Recent research, however, suggests that previously unexplored marine habitats, especially the deep sea and the ocean floor, may harbor millions of additional species, thus rivaling the species richness even of the tropical forests. Moreover, if we measure diversity in the broader taxonomic categories—phyla, classes, divisions—then the greatest variety of life on Earth is unquestionably contained within the seas (Thorne-Miller and Cantena, 1991). It is not uncommon to find representatives of a dozen or more basic classes or divisions in the same small space—a breadth of diversity that has no match on land.

Fish, marine mammals, mollusks, and corals are the best-known groups of marine organisms. However, major groups of organisms and new habitats are still being discovered. The phylum Loricifera was first described in 1983 (Kristensen, 1983), and an entirely new habitat was revealed with the discovery of ocean vent systems. The bottom of the ocean is still largely unexplored; assaying and understanding its biological diversity will require resources equivalent to those committed for exploring the Moon. Because such research depends on costly and specialized equipment, funding for ships and associated sampling tools is a limiting factor (NSB, 1989).

The importance of marine biodiversity is almost as vast as the oceans themselves. Much of the Earth's human population depends on the oceans, especially marine coastal systems, for food. In the developing nations, more than half of the population obtains at least 40 percent of its animal protein from fish (WRI, 1986). Some 9,000 species of fish are currently exploited for food, although only 22 are harvested in significant quantities on a global scale (WRI, 1987). Approximately 80 percent of the marine species of commercial importance occur within 200 miles of a coast. Marine flora and fauna are also extensively used in the production of antibiotics and other pharmaceuticals, food additives and processing agents, and a variety of manufactured goods.

Above and beyond these commodity values, marine organisms are critical determinants of the structure and function of the global ecosystem. Marine phytoplankton, for example, are the foundation of marine food chains and play an important role in atmospheric dynamics. The interactions among marine biota, the Earth's geochemical cycles, and global climate change are just coming to light, and even our most advanced computer models have been able to offer only the roughest approximations of the feedback mechanisms involved in the maintenance of biospheric conditions. The study of marine biodiversity is thus critical to understanding environmental dynamics on the global, as well as on local and regional, scales.

Interest in the conservation of marine biodiversity is a relatively recent phenomenon. The immensity that makes oceans such a challenge to study has also made it possible to believe that anthropogenic disturbances would remain limited in their environmental impact. Compared to terrestrial environments, oceans provide relatively stable, extensive, open, well-buffered habitats for the organisms that inhabit them. Nonetheless, the threats to marine diversity are much the same as on land: habitat destruction (especially in coastal, estuarine, wetland, and coral reef systems); pollution (including suspended sediments, nutrients, and toxics); overexploitation of harvestable species (including fish, shellfish, turtles, and mammalian species); and the specter of global climate change with all its attendant marine impacts (Soulé, 1991; Thorne-Miller and Cantena, 1991).

Although the biota of oceans has been protected from many of these impacts by the extent of the medium itself, environmental stresses can be expected to place the same pressures on marine systems that they are placing on terrestrial systems. So little is known about marine biota that rates of extinction are difficult to estimate. Ray (1988), however, suggests that the degradation of coastal zones is occurring as rapidly as tropical forest destruction, and recent findings indicate that coral reefs may be among those communities most seriously imperiled by human activities (Salvat, 1987; Guzman, 1991). As in terrestrial systems, inventories and ecological studies are needed for all oceans, with special emphasis on those habitats most immediately threatened.

This brief review does not reflect the full status of scientific knowledge with regard to specific taxa, geographic areas, ecosystems, or habitats, and only touches on genetic-level diversity and the vitally important relationship between ecosystem dynamics and diversity. As we seek the means to slow or reverse the losses, we will have to secure increased support for established scientific efforts in systematics and resource management, and for relatively new scientific endeavors in such integrative, applied fields as sustainable agriculture, conservation biology, and restoration ecology. We face an unprecedented situation that demands new combinations of the basic and applied sciences, the expertise of specialists and the vision of generalists, conceptual clarity as well as concrete experience. The science of biological diversity and its conservation demands not only more knowledge but new kinds of knowledge, and new ways of synthesizing what we know.

IMPLICATIONS FOR DEVELOPMENT AGENCIES

Biological diversity reaches its highest levels, and faces its greatest risks, in the developing nations of the world, primarily because of intensive resource exploitation and the extensive alteration of habitats. This is due in part, however, to international markets, development policies, and lending practices that transfer financial resources from developing countries to industrial countries and undermine the capacity of developing countries to sustainably manage their resources.

Rapid population growth, extreme and persistent poverty, social inequity, institutional breakdown, and perverse policy incentives have brought unstable economic conditions to many developing nations. In response, many of these countries have had to adopt short-term development agendas and exploitative resource management practices aimed at increasing foreign exchange earnings from their undiversified economies. Trade in elephant ivory (mostly illegal) and tropical timber (legal) provides obvious examples that have important consequences for the maintenance of biodiversity, but other less publicized practices—overgrazing of ranges, expansion of cash crop agriculture, intensified shifting cultivation—also lead directly to the demise of species and habitats.

As a result of these interrelated social, economic, and environmental trends, many developing countries have begun to question the sustainability of current resource management practices and look for more promising alternatives. The policies and funding practices of international development agencies, if directed toward wise, long-term commitments of assistance, can aid in this by affording developing countries greater economic stability and hence greater national capacity to preserve biological diversity. In the past, development agencies have funded infrastructural development activities, agricultural expansion programs, dams, and other large-scale projects that have contributed directly to the loss of biological diversity, while doing little to ease the indirect causes of resource decline (NSB, 1989). A new vision is necessary at all levels of the development community—one that recognizes the inextricably connected fate of human communities and the biotic community, of development and conservation.

Biological diversity is, in the most literal sense, the basis of sustainable development and resource management. By conserving biodiversity, we retain not only plants and animals, soils and waters, but the foundations of sustainable societies and the availability of options for future generations. Fuelwood gathering, to cite just one example, is a significant contributing factor behind the rising rates of deforestation in many parts of the tropics. A billion and half people in developing countries depend on firewood as their major fuel source. In many areas, expanding demand and declining local supplies have led to excessive harvest rates, and acute fuelwood shortages, and subsequent decline in soil and water resources. Developing renewable, cost-effective alternative energy sources, sustainable agroforestry systems, and more productive sources of firewood, charcoal, and timber will require greater attention to potentially useful species and genetic resources (NRC, 1991a).

Biodiversity, in short, must come to be seen as an inherently important aspect of every nation's heritage and as a productive, sustainable resource upon which we all depend for our present and future welfare. The conservation of biological diversity is not merely an obscure, hitherto neglected area of endeavor whose importance has only now been discovered; rather, it is a fundamental concern that has been absent in short-term development planning, at the risk of long-term social and economic well-being.

Responding to Research Needs

In both the developing and the developed nations, immediate action needs to be taken to protect biodiversity. At the same time, there is a continuing need for research on biodiversity that improves our knowledge base and our management capacities, and leads to the development of new ways for people to live with, and not at the expense of, their biological resources.

It is unlikely that poor countries will be able to support major biodiversity research enterprises, however important, in the near future. If global environmental and scientific objectives are to be served, more effective means for north-south transfers of funding must be found, and more productive mechanisms for scientific collaboration must be invented (NSB, 1989). The international development agencies are essential in this regard. Other organizations are unlikely or unable to provide the necessary funds. In the long run, this assistance will allow developing nations to move toward greater independence by strengthening in-country research institutions. As their research capacity increases, so too will their ability to chart their own course of sustainable development.

As they seek to meet these growing research needs, development agencies will themselves have to undertake institutional changes. Research on biological diversity is necessarily broad based and multidisciplinary, and the administration of research within the agencies must reflect this. Overlapping areas of biology, including ecology, sustainable agriculture, and conservation biology, are critically important in addressing the needs of developing countries and must be given greater support. More support must also be given to research that integrates economics, the social sciences, and biodiversity conservation. Above all, research must be carried out largely by people in and of the countries involved.

Long-term institutional commitment is necessary. Support for these changes must be incorporated wherever possible into the human resource development programs of technical assistance agencies. All personnel should be given training in biodiversity science and policy. More personnel with the requisite background knowledge must be brought into the agencies on a permanent basis and given adequate specific training, as well as opportunities to remain up to date on research in their fields. Although development and science agencies can play a leading role in promoting these efforts, their work must involve agencies, institutions, and organizations that have not traditionally taken part in conservation activities. Finally, development agencies must have a "built-in" capacity to review outcomes, monitor practices, and recommend adjustments in policies that affect the status of biological diversity.

Several development agency research programs have begun to reflect these needs. The U.S. Agency for International Development, for example, provides funds for innovative research on biodiversity under its Program of Scientific and Technical Cooperation (PSTC) and its Sustainable Agriculture and Natural Resource Management (SANREM) Collaborative Research Support Program. Support for this kind of research should be expanded and strengthened. Agencies will need to find creative ways to sustain funding for these endeavors over many years, even indefinitely. National biological inventories, for example, could well be funded by pooling the resources of all international assistance agencies functioning within a given country.

The research agenda outlined in the remainder of this report is intended to assist development agencies in their efforts to respond to these research needs. Research cannot, in and of itself, conserve biodiversity in developing nations any more than it can in the developed nations. What research can do, however, is provide the people and the leaders of these nations with information that may help them to improve their lives, while securing the biological legacy on which their livelihood depends.

2
Biological Aspects Of Conservation

In the past, national and international development agencies have seldom relied on—or called for—basic information on biological diversity. This can no longer be the case. Many development projects include a significant natural resource component and thus require sober analysis of their environmental impacts. More broadly, international agencies and resource and planning ministries in developing countries need information about biological diversity to formulate development plans and specific projects that are both successful and sustainable.

Pertinent information on biological diversity in most developing countries is too sparse or scattered to be of practical use. Often it is unavailable altogether. A good deal of "gray" literature exists—unpublished reports, files in government archives, studies of limited distribution. The most important of these should be analyzed and made more accessible. In general, however, the required information can be gathered and disseminated only through systematic efforts to strengthen the entire research process.

Development agencies need to know which kinds of research are of greatest relevance as they assist client governments and develop the rationale to secure funding for this research. A large and growing body of literature describes conservation strategies appropriate to different species, ecosystems, and regions in developing countries. This includes journals such as *Biotropica*, *Biological Conservation*, and *Conservation Biology*. Recent agendas, involving a range of basic and applied research needs, can be found in *Research Priorities in Conservation Biology* (Soulé and Kohm, 1989); *From Genes to Ecosystems: A Research Agenda For Biodiversity* (Solbrig, 1991); and *The Sustainable Biosphere Initiative: An Ecological Research Agenda* (ESA, 1991). Subsequent chapters of this report focus on the socioeconomic and cultural aspects of biodiversity research in developing countries. This chapter provides an agenda for biological research that must be undertaken to provide a sound foundation for these human dimensions of successful conservation.

The state of knowledge of biological diversity, described in the previous chapter, suggests that the most basic research requirement is to gain a more complete sense of "what's out there." The committee that produced the 1980 National Academy of Sciences report *Research Priorities in Tropical Biology* recognized this fundamental need and called for a "greatly accelerated . . . international effort in completing an inventory of tropical organisms" (NAS, 1980). Although these efforts have accelerated to a degree, the task has become far more urgent, complex, and challenging in the interim.

Effective conservation of biological diversity requires more than just basic knowledge of its components. We need to know as well the distribution of biological diversity and those areas where it is most concentrated. We need to know the potential benefits that organisms can offer to humanity and, at least in a general way, how they and the biotic communities they form are faring. We need to understand better the ecological dynamics of the systems in which organisms exist, the temporal and spatial patterns that govern their fate, and the best means to conserve both organisms and habitats over the long run. We need to develop methods to use biological resources without depleting them or undermining the human communities with which they coexist. Finally, we need to learn better how to restore those lands and waters that have been degraded by unwise development.

The challenge of biodiversity research entails not only the gathering of information but its management, application, and communication. Likewise, the quality of research depends upon the people and institutions who perform it. These considerations are especially important in the developing nations of the world, and are addressed as part of this research agenda. The specific recommendations offered flow from the general conviction that the comprehension and conservation of biodiversity in developing nations represent a challenge of such magnitude that all links in the chain of research and application must be strengthened to ensure success.

These recommendations have been formulated with the understanding that many development agencies have central (global or worldwide) interests as well as country-specific programs. Research and related activities appropriate for both have been included. In general, to have the greatest immediate as well as long-term impact, centrally funded research should be conducted in concert with in-country activities, even when problems are addressed on a global scale. For example, research on salt-tolerant plants that can restore saline soil to agricultural productivity should be undertaken in a location where this is a problem, even if the work is centrally funded and conducted in collaboration with U.S. investigators. Furthermore, centrally funded research is likely to be more basic in nature, and linking it to in-country projects can demonstrate to agency personnel how basic research is directly applicable to development activities.

BIOLOGICAL SURVEYS, INVENTORY, AND MONITORING

Successful, long-term conservation of an area or ecosystem relies on knowledge of its biological diversity coupled with integrated efforts to protect and manage that diversity in a sustainable manner. One of the first steps in this process is to ascertain its fundamental biological characteristics: the genetic strains, species, and ecological assemblages present; their distribution, abundance, and patterns in the landscape; their role in ecological processes; their proven or potential utility for human benefit; and trends in their status as a result of human or natural disturbances. Full understanding of biological diversity, even in a small area, is a task requiring decades, if not centuries, of intensive research. Biological surveys, inventories, and monitoring can, however, provide the basic knowledge required to enhance local scientific and technical expertise and to initiate sound conservation strategies.

Biological surveys, focusing on species diversity, are necessary on both national and global scales. National biological inventories provide a finer-grained view of biological diversity and can be used to establish national conservation programs and policies, whereas a global survey will provide much needed information on the extent, distribution, status, and fate of biodiversity worldwide. These efforts can serve not only to tell us the status of biodiversity, but to identify valuable biological resources, some of which are unknown, while others are locally known but have potential for much wider use. Many plants of current or potential commercial value (e.g., the maize *Zea diploperennis* and the tomato *Solanum pimpinellifolium*, both recently collected from Mexico) were discovered in the course of routine plant surveys. Inventories and surveys also provide baseline data against which to monitor changes in biological diversity and to trace the environmental impacts of development projects.

For all groups of organisms, sampling those that occur in threatened regions is of special importance because natural communities are being altered or destroyed so rapidly. Large numbers of endemic species are being lost in the world's critical centers of endemism, or "hot spots" (Myers, 1988). It should be emphasized that we do not have the slightest idea how many species of nematodes, mites, and many other taxonomic groups exist within or beyond these hot spots, nor do we know if other hotspots exist. If we are ever to know, we will have to sample these groups and areas. In cases where immediate information on an area's species diversity is needed, new rapid assessment methods may be required (Roberts, 1991). Particularly in species-rich areas—but throughout the developing nations and in threatened habitats worldwide—inventory and preservation are of immediate and critical importance.

To achieve an acceptable standard of knowledge about the diversity of the world's biota, the following actions are needed.

National Biological Inventories

National biological inventories should be organized, funded, and strengthened in each country of the world.

This should be the top priority for development agencies in biodiversity research. National inventories offer exceptional possibilities for professional linkages and community development and provide the thorough knowledge of organisms necessary for intelligent management of biological diversity to solve any number of practical problems. In many cases this work can, with appropriate investments, be implemented through existing institutions, but should be coordinated through the establishment of national biological institutes (or equivalent centers) such as Costa Rica's Instituto Nacional de Biodiversidad (see sidebar).

Information gathered in national inventories and stored in data banks provides a foundation for sustainable economic development and is important both in the formulation, preservation, and management of natural areas and in the design of improved agroecosystems. The establishment of national biological inventories, in fact, would entail the implementation of many of the other recommendations offered in this chapter and would serve as an integrating force and focus for research on biological diversity and its management.

Given the scarcity of biodiversity data for most ecosystems in developing countries, the worldwide shortage of trained personnel, and an almost total lack of local taxonomic expertise, it is a substantial challenge to survey a country, protected reserve, or even a small potential project site. For this reason it is recommended that surveying activities concentrate on the biological groups that are best known as well as ecologically and economically important, primarily higher plants and vertebrates. Focusing on these groups would allow substantial short-term progress to be made in what must, of necessity, be a long-term, sustained effort.

Local scientists working in national herbaria, museums, zoos, aquaria, arboreta, and universities should be responsible for these activities if possible. The surveys should involve strengthening these institutions and training technicians, parataxonomists, and graduate students. If local expertise is not available, local scientists should be trained by foreign experts invited to collaborate in the surveys. In the

INBio

Costa Rica's efforts to inventory and manage biological diversity are coordinated through the Instituto Nacional de Biodiversidad (INBio). The inventory work at INBio is conducted largely by a "small army" of lay persons trained for the task, called "parataxonomists," who work in close collaboration with national and international curators and professional taxonomists. They are recruited from many sectors of the Costa Rican population. INBio organizes a five-month course in alternate years that includes basic information about the taxonomy, biology, and ecology of plants and insects, as well as basic instruction in collecting and curating techniques. This results in a large volume of partially identified specimens, and provides an informal but important means of information exchange. As they return to their home communities with an appreciation of the value of biodiversity, parataxonomists continue to be paid for their work in the field, bringing indigenous knowledge back to INBio. The parataxonomists also constitute a pool of individuals who may, and in many cases do, elect to obtain further training.

INBio has also developed data bases to house and organize the large quantities of information collected daily. This illustrates yet another important aspect of INBio's work. While undertaking the national inventory of biodiversity and providing a focal point for its management, INBio also puts the knowledge of Costa Rica's biodiversity to work for its people. By developing the commercial potential of its biotic resources through partnerships with industry, Costa Rica can ensure that conservation activities support themselves, as well as the people of Costa Rica (Tangley, 1990).

training of specialists, links with agencies and universities in the developed countries will be important. All surveys should be conducted by multidisciplinary teams able both to assess biological diversity and to describe the physical and socioeconomic characteristics of the area surveyed.

In practice it is often impossible even to recognize the numbers of species present in a given sample without having a specialist's knowledge of that particular group. For that reason, both monographic studies, which constitute the principal activity of many systematic biologists, and regional inventories are of primary importance. Because the shortage of trained personnel will affect not only the conduct of surveys but all aspects of conservation research, it is imperative that new rapid-assessment survey methods be developed and that intensive training courses be given on these methods, within country if possible, or at least regionally.

The development of computerized data bases, which can provide information about organisms rapidly and efficiently on a regional, national, or global basis, is extremely important. Such data bases will be of use to a wide range of agricultural scientists, biologists, resource managers, environmental engineers, farmers, teachers, and others (Morin et al., 1989). They can be continually updated and corrected and, as countries bring them into operation, can simplify the coordination of surveys, providing a regional picture of biological diversity. These national and regional efforts, building on existing Conservation Data Centers in other countries, can well add up to a global strategy.

In addition to the research priorities suggested here, priorities in methodology, training, and institution building are suggested in other sections of this chapter.

Research Priorities

- Undertake biological surveys in all national parks, preserves, and conservation units, either as an integral part of conservation projects being run in those units or as part of a national natural resource assessment. Countries should determine which taxa are most important to survey.
- Employ statistical methods in surveys to estimate the abundance of species present or completeness of sampling.
- Conduct surveys on wild relatives of commercial domesticated and semidomesticated crop and animal species as part of all agricultural improvement programs.
- Conduct surveys of at least the higher plants and vertebrates for all agroforestry, diversified agriculture, sustainable agriculture, animal improvement, animal introduction, fisheries, and forestry project sites.
- Conduct initial surveys of aquatic flora and fauna to serve as baseline data for management, long-term monitoring, and project impact assessment.
- Monitor changes in species diversity in project sites during and after the project for a period of 10 years.
- Determine which species should be reintroduced and where populations can be self-sustaining.
- Evaluate photographic methods for rapid taxonomic surveys and associations in ecosystems.

• Develop systems (e.g., using microcomputer programs for numerical keying of species characteristics and, ultimately, CD-ROM discs with video color images) to facilitate the identification of material collected during surveys. Development of these systems will benefit from the widespread availability of personal computers and computer "literacy" in most developing countries in recent years (NRC, 1989). These systems will also promote collaboration among scientists in industrial and developing countries.

• Develop identification manuals in local languages with workable keys. Include information geared to the general public and to users in different professions.

Global Biological Survey

A strategy for gauging the magnitude and patterns of distribution of biological diversity on Earth should be coordinated and implemented.

A global survey, drawing on the work of national biological inventories and supplemented by extensive surveys of particular localities, should be undertaken immediately. The National Academy of Sciences study *Research Priorities in Tropical Biology* recommended that a comprehensive, multidisciplinary survey of well-known groups of tropical organisms (e.g., plants, vertebrates, and butterflies) be undertaken to provide an index to the patterns of distribution and the nature of communities throughout the tropics (NAS, 1980). This recommendation is even more timely now. Such a survey would not only serve as an index to diversity in the tropics, but would provide the cornerstone for a global-scale effort.

This task cannot be postponed. The rapid decline of species of all groups and the deterioration of entire ecosystems are occurring so rapidly that even the possibility of a reasonably complete assessment will slip from our grasp if not undertaken in the next two decades. A global survey would give our estimates of diversity a stronger foundation, and help us understand all that we stand to gain or lose in this critical period.

The initial focus of a global survey should be plants, vertebrates, butterflies, and a few others groups of organisms that are well known, well studied, or of particular economic importance (e.g., mosquitoes). Inventories of these high-profile groups, as noted, can serve as indexes to entire areas and as indicators of basic biogeographical patterns. Lesser-known groups can then be sampled and eventually more completely inventoried to supplement this information, providing additional insight into these broad patterns and the ways in which they are changing.

For groups such as vertebrates, detailed study of all species present in particular localities is the only way to learn about their abundance and distribution. Lesser-known but ecologically significant groups of organisms—insects, free-living nematodes, ciliates, mites, fungi, bacteria—deserve special attention. Present ecological and systematic knowledge of these groups is very limited. From what little we do know, it is clear that hundreds of thousands or even millions of species still remain to be identified and described. For these groups, research should focus on the development of new taxonomic techniques, and funds should be provided for training and employment opportunities.

At the same time, comprehensive estimates of the number of species present in representative and threatened areas of the tropics and elsewhere must be undertaken (May, 1988; NSB, 1989; Gaston, 1991; Erwin, 1991). Study locations should be chosen carefully in order to devote sufficient effort to areas that are poorly known, species-rich, under imminent threat of destruction, and unique in their habitats and genetic stocks, such as Madagascar, island groups, and the Chilean Andes (see NAS, 1980; Myers, 1988; NSB, 1989). They must be organized so as to concentrate resources of time and expertise, and avoid dilution of effort. Coordination with national biological inventories is critical.

These surveys, conducted at selected study sites, could begin to provide a sound basis for estimating the abundance and diversity of organisms on Earth. For example, the study of 200 scattered locations in great depth, or of more locations in less depth, could provide a vastly improved empirical basis for estimates. Already protected areas would, of course, be prime candidates for study sites, but even 200 locations would represent only a portion of those currently under some type of protection; the number of locations at risk is much higher (NSB, 1989).

Coupling this with our knowledge of the abundance and distribution of organisms in already well-studied areas (such as La Selva in Costa Rica or Barro Colorado Island in Panama) will likewise be very important. By studying groups of organisms in areas where ecosystem functions such as the flow of energy and the cycling of mineral nutrients are understood, we will be able to evaluate better the characteristic ecological roles of those groups and use this information to arrive at some approximation of biological diversity under similar ecological conditions (May, 1988). This will also allow us to understand better the ecological basis of biodiversity and its functional significance within ecosystems (ESA, 1991).

Marine biodiversity must also be considered much more prominently in the context of global biodiversity research. The diversity of marine systems and organisms is often overlooked, in large part because the marine environment is extensive, difficult to study, and mistakenly assumed to have a limitless capacity to absorb human impacts. Global marine inventories would provide information about the processes that regulate biodiversity in marine communities and the impact of pollution and harvesting on these processes. In addition, the basic biological structure and the component species of marine systems are poorly known. The distribution and migratory patterns of most species remain unknown, and many benthic organisms remain undescribed. Accurate assessments of human impacts on these systems requires this baseline data. Traditional methodologies coupled with modern technologies (e.g., manned submersibles, unmanned remotely operated vehicles, and color-scanning of data to assess phytoplankton distribution over space and time) can, with appropriate support, make great strides in improving our understanding of marine biodiversity. These tasks could be undertaken on a regional basis (e.g., through the Regional Seas Program of the United Nations Environmental Programme), and at the continental and global scales.

Research Priorities

- Initial efforts to coordinate a strategy for a global biodiversity survey should be supported. These should included regional and global conferences to reach agreement on methodologies and institutional linkages.
- Opportunities for global cooperation should be taken into account in the establishment and long-term planning of national biodiversity inventories.
- Support should be given to comprehensive surveys of localities and groups of organisms that can serve as indices and indicators of basic biogeographical patterns.
- Develop inventory and training techniques to allow increased taxonomic work on nematodes, ciliates, mites, bacteria, and other little-known taxa.
- Much more attention should be devoted to research on the diversity of marine communities and organisms.

Screening of Organisms

The screening of plants, animals, fungi, and microorganisms for features of potential human benefit should be systematized and accelerated.

National biological inventories should provide screening opportunities for new natural products and rational methods for using materials derived from them. The tropics alone contain more than 170,000 species of flowering plants. Although fewer than 1 percent have been examined by modern medicine for utility or commercially exploited, even this small fraction has yielded a variety of useful drugs, food crops, biomass stocks, latex, oils, resins, fuels, and other products. Many have become major international commodities and the base of national economies. Furthermore, most people in the world depend on these plant products for medicine, fuel, and other goods.

A wider understanding of biological resources, together with their appropriate use, can greatly enhance human welfare. At the same time, the very process of examining a nation's biological diversity for its economic potential can focus attention in a manner that advances broader conservation goals. International agencies charged with improving the status of people in developing countries have a singular opportunity to help these people realize that potential and to increase local understanding, preservation, and application of traditional resource uses. Screening research is essential in this endeavor.

For what qualities and products should organisms be screened, and how should this type of research be integrated into the activities of development agencies? One part of a screening program should be directed toward finding products that can help solve problems unique to developing nations. In many developing countries, for example, traditional therapeutics can be used to treat diseases in communities that lack access to modern medicine and are rapidly losing access to traditional cures. Pharmaceutical compounds and less processed plant or animal extracts are particularly important in this regard. Experience suggests that developing nations themselves will have to undertake the majority of this type of research or create incentives for the involvement of pharmaceutical companies in the developed nations. These might involve, for example, the trading of raw materials from the tropics for screening as possible treatments for cancer, AIDS, and other diseases, in exchange for which the pharmaceutical companies would agree to screen for treatments for tropical maladies. Studies of traditional cures available in local markets and consultation with individuals locally trained in traditional skills are also important steps toward managing biodiversity for the improvement of health care.

The screening process should also serve to identify organisms of potential benefit in agriculture and in the provision of environmental services. These include trees and alternative crops for use in agroforestry, mixed cropping, and other sustainable agroecosystems; effective biological control agents for agricultural insect and plant pests; and plants (especially legumes) of special importance for erosion control and land rehabilitation (NRC, 1975; 1979). Traditional food and fiber crops should also be screened to identify those that might be developed into national or international market commodities.

Genetic engineers are already beginning to take advantage of screening to identify and transfer useful genes from plants and animals. The variety of habitats in the tropics and subtropics, many of which are highly endangered, constitutes a veritable storehouse of genetic resources, exceeding anything that Temperate Zone experience suggests. The accelerated loss of individual species in the tropics thus also implies the loss of an unimaginably vast diversity of genetic structures (NRC, 1991a; 1991b).

Candidates for new products should first be sought among the organisms known and used by indigenous peoples. If indigenous knowledge has not been documented and compiled, doing so should be a research priority of the highest order. Indigenous knowledge is being lost at an unprecedented rate, and its preservation, preferably in data base form, must take place as quickly as possible.

Taxonomists, chemists, biochemists, anthropologists, ethnobiologists, and other scientists need to be trained to work in multidisciplinary screening teams (see Chapter 4). Beginning with interviews of local people regarding their use and management of local resources, these teams should be able to obtain information on the indigenous knowledge base, identify plants and animals for screening, and conduct screening studies. Resources must also be devoted to strengthening institutions at which such work is currently being done. Finally, because we are severely limited by our inability to quickly survey species-rich tropical areas, resources should be provided to develop improved bioassays and rapid screening methods.

If possible, screening teams should be composed of local scientists. If this is not possible, the teams should work closely with local scientists and establish partnerships with local universities, agencies, nongovernmental organizations, and other institutions. In addition, this research will require a deep consideration of intellectual property rights. Legal opinions differ, but many patent experts agree that when indigenous peoples reveal their uses of plants and animals, their exclusive right to that information is lost, just as trade secrets enter the public domain when they are revealed. This complicated legal and moral issue can be dealt with only through careful planning and the negotiation of contracts guaranteeing equitable distribution of royalties to indigenous groups. The National Cancer Institute of the National Institutes of Health, Cultural Survival, and the Institute of Economic Botany of the New York Botanical Garden have formulated agreements that can serve as models for researchers.

Research Priorities

- Identify tropical diseases for which (1) drug treatments are unknown or of limited effectiveness, (2) treatment is expensive and beyond the reach of most of the population, and (3) local remedies exist.

• Screen local remedies for active compounds and determine how similar these are to existing drug formulations; identify opportunities for developing low-cost, standardized, effective treatments for curable ailments.

• Examine indigenous cropping systems for biocontrol methods and agents; determine the efficacy of local means of biocontrol compared to agricultural techniques dependent on purchased inputs.

• Study the relationship between the concentration of active compounds within different populations of species and the environmental conditions under which these populations occur.

• Compare the qualities of local sources of waxes, oils, and other products to those of commercially available industrial and domestic brands.

• Develop means to ensure that intellectual property rights are fully considered and secured during the screening process and subsequent development of products.

Monitoring

To detect, measure, and assess changes in the status of biological diversity, appropriate monitoring methods employing specific indicators of biodiversity attributes should be implemented.

National biological inventories and a global-scale survey will provide basic information about the degree and distribution of species diversity, while screening allows us to determine more systematically the current and potential uses of organisms for human needs. Such efforts must be complemented by the development and implementation of monitoring methods that track the continually changing status of biological diversity.

Because diversity is characteristic of life at multiple levels of organization—genetic, population, species, community or ecosystem, landscape or region, and global—monitoring presents almost incomprehensible challenges. Obviously, the changes that take place at all these levels cannot be constantly or completely measured, nor is it necessary to do so. An effective monitoring program, however, should recognize that various levels of biological organization do exist; that although they are distinguishable from one another, they are not disjunct; that, instead, they are "nested" within and affect one another; and that, in monitoring, different levels of resolution will be necessary to address different scientific and management-related questions (Allen and Starr, 1982; O'Neill et al., 1986; Noss, 1990). For example, tracing the impacts of acid precipitation on biological diversity requires an understanding of regional biogeographic patterns related to topographic and climatic variation and soil variations at the landscape level; patterns of species richness within terrestrial and aquatic systems; the differential physiological responses of individual species to variations in acidity; and intraspecific genetic responses to selection pressures associated with acidification.

In practice, this hierarchic approach implies that a variety of indicators of biological diversity, and hence monitoring tools, must be employed (Noss, 1990). Many of these tools—Geographic Information Systems (GIS), other remote sensing data, landscape pattern indices, ground-level plant community sampling methods and animal censuses, and electrophoresis, to name a few—have long been in use, although their application to questions of biological diversity may be relatively recent. The key to successful monitoring programs is to maintain the ability to detect general changes in the status of biodiversity, to choose those tools that are adequate and appropriate for particular conservation problems, and to coordinate their use so as to test hypotheses of broad relevance to conservation.

A general guideline is to proceed "from the top down," beginning with a coarse-scale inventory of landscape pattern, vegetation, habitat structure, and species distributions, and then overlaying data on environmental stresses to identify biologically significant areas at high risk of impoverishment (Noss, 1990). Whereas extensive methods of monitoring (e.g., of forest cover) should proceed at the total landscape level, more intensive research and monitoring methods should be applied to high-risk ecosystems and to lower-order components of biological diversity (e.g., rare plant communities, communities rich in endemic species, endangered species, and disjunct populations).

Monitoring should not be undertaken as a separate activity, but integrated into other research activities. National and global inventories should monitor changes in ecosystem and species diversity during initial investigations and periodically thereafter. These inventories, in turn, should suggest genetic resources, species, and habitats that require more intensive monitoring. Long-term, site-specific ecological research (see below) should have, as one of its primary responsibilities, the monitoring of biotic diversity as a function of interspecific relationships (e.g., between native and introduced species), habitat variables (e.g., canopy openness in forests), and abiotic factors (e.g., disturbance regimes). Most research projects involving the application of conservation biology principles—for example, the design of reserves, the biology of rare and declining species, and the role of keystone species—should involve monitoring as a part of the research methodology.

In designing research projects, thought should be given to the manner in which results can be applied to other monitoring efforts. Conservation

Data Centers, computer inventories, and other information management technologies and networks should develop the capacity to receive and distribute data for monitoring purposes. The training of personnel should include instruction in monitoring methods, while the strengthening of in-country institutions should include an improved capacity to carry on monitoring programs.

In supporting enhanced monitoring programs, international development agencies should coordinate their work with that of the United Nations Food and Agriculture Organization, the World Conservation Monitoring Centre (WCMC), and other organizations. The WCMC, for example, was established jointly by the International Union for the Conservation of Nature, the United Nations Environmental Programme, and the World Wide Fund for Nature to collect data on threatened, endangered, and very rare species in protected areas of the world, and to monitor the international trade in threatened species and wildlife products (McNeely et al., 1990). As expanded, longer-term biodiversity monitoring efforts become increasingly important, this kind of coordination will be necessary to avoid duplication of effort in what is already a demanding task.

Research Priorities

- Develop appropriate statistical analysis techniques for monitoring projects and validate the effectiveness of indicators for use in monitoring.
- Explore and demonstrate the capabilities and limitations of remote sensing, satellite imagery, aerial photography, and other more experimental techniques for monitoring. What are appropriate levels of analysis for each? Can these technologies, which have been applied mainly in Temperate Zone countries, be transferred easily to tropical ecosystems?
- Work with geographers to create up-to-date, well-documented vegetation maps based on detailed field studies that can serve as the basis for other monitoring methods. Document the status of vegetation and point out areas requiring further study.
- Determine which ground-level plant and animal sampling methods best indicate the diversity of tropical ecosystems and are most capable of incorporation into monitoring programs. This should involve the adaptation to tropical systems of methods used in temperate zones, as well as the development of new methods especially suited for these regions.
- Incorporate more monitoring responsibilities into long-term site-specific ecological research. These extended and detailed studies can

provide much of the information necessary to interpret shorter-term and necessarily less detailed monitoring studies.

CONSERVATION RESEARCH

Gathering basic information on biological diversity, through the means recommended above, must be a constant effort. In the meantime, the best possible conservation measures must be undertaken, especially where diversity is at greatest risk. These efforts also require research.

Conservation, in this context, should not be construed to mean either strict preservation or intrusive management, but these measures and all others that can protect and restore the biological diversity inherent in an area while improving the long-term well-being of the people who live there. This entails a spectrum of appropriate land uses, from parks and natural areas to extractive reserves to sustainably managed agroecosystems. In this effort, it is important to appreciate the different spatial scales on which different landuses operate; to understand them in their landscape, regional, and even global contexts; and to coordinate them so as to conserve their full range of values.

This is a massive challenge, and the role of the international development community is critical. Development agencies are able to provide the institutional and financial support necessary for both immediate and long-term research. They have access to expertise in the wide range of relevant fields and disciplines that must take part in the research. They have a record of accomplishment in collaborative research in many sectors (particularly agriculture) that can be adapted to research on the conservation of biological diversity. They are also able to coordinate research among developing and developed countries, among countries within a region, and among regions around the world. All of these features are vital to advances in conservation research and the application of findings in effective programs.

Research on conservation must integrate and extend the basic information gathered through biological surveys and inventories to increase our understanding of ecological dynamics in different systems and regions. Such understanding can be gained only through long-term studies of specific sites. These regions and systems should also be the focus for studies of basic concepts and emerging principles in conservation biology (including, for example, the optimal design of conservation reserves and buffer zones, the impacts of habitat fragmentation, and the involvement of local communities in conservation programs). Strategies for sustainable use of land and biological resources, and for returning something of the value of biodiversity to developing countries, require greater scientific scrutiny. Finally, research on the restoration and utilization of degraded lands must become a much higher priority.

To advance our understanding of successful conservation strategies and methods, the following actions are needed.

Site-Specific Research

To advance the understanding of ecosystem composition, structure, and function; to use this knowledge to link basic and applied research, sustainable land use and development, and the conservation of biological diversity; and to provide baseline data for environmental monitoring, support should be given to long-term ecological research at selected sites in developing nations.

Progress toward truly sustainable land use systems requires information on the effect of management options on ecosystem dynamics, and this information can be gained only through long-term research. Long-term basic and applied research is especially needed in the tropical ecosystems of the developing world, where few comprehensive investigations of ecosystems and land use have been undertaken. Development has proceeded essentially by trial and error, often with disastrous consequences. Without information on the effects of management decisions on ecosystems, and baseline data against which to measure these effects, development efforts will often continue to result in the loss of biological diversity and the degradation of other natural resources. The best way to develop this information base, avoid duplication of effort, integrate experimental design and results, advance our understanding of environmental monitoring, and focus on applied conservation and land use methods is to coordinate research at several selected sites.

Long-term research at specific sites has provided important insights into temperate ecosystems. In the United States, for example, studies of northeastern forests at the Hubbard Brook Experimental Forest in New Hampshire have investigated the effects of different harvesting strategies on ecosystem processes (Bormann and Likens, 1979). Similar research projects are focusing on North American ecosystems as diverse as coastal estuaries, shortgrass prairies, coniferous forests, and arctic tundra. The Long-Term Ecological Research (LTER) Program of the U.S. National Science Foundation works to support and coordinate 17 of these long-term research sites, providing a network of ecological investigations for comparative analysis (Franklin et al., 1990).

There has been movement toward establishing an international research network along similar lines (Risser and Melillo, 1991). The United Nations-sponsored Man and the Biosphere (MAB) Project 8, which focuses on the development of a worldwide network of protected areas for long-term ecological research and monitoring, has many of the same goals (Unesco, 1974). Although much work has been done over many years at, for example, Barro Colorado Island and La Selva, very little truly long-term research has even been initiated in the tropics.

The international development community must play a leading role in supporting this type of research. As experience has shown, international development agencies can no longer afford not to have information on the long-term environmental consequences of development projects. Because they provide the resources for major management interventions in developing countries, they are the appropriate sponsors of studies assessing the effects of those interventions. Mechanisms to support continuing, site-specific, multidisciplinary studies should be built into the programs and funding procedures of development agencies.

In the past, development agencies have failed to undertake long-term studies for a variety of reasons. The typical two- to five-year project administration cycle has not been conducive to long-term planning or oversight. Funding procedures provided support for project installation and program implementation, but not for subsequent monitoring or environmental assessment. Long-term ecosystem monitoring was assumed to be beyond the agencies' mandates. In the future, to accomplish their mission, development agencies will need information that only long-term, site-specific research can provide.

Research Requirements

Until recently, comprehensive ecosystem studies were comparatively difficult, time consuming, and expensive. The techniques for quantifying interactions among ecosystem components—for example, nutrient movement between soil, plants, microorganisms, and animals—were primitive and labor intensive. Recently, however, the development of portable equipment for sampling and biochemical analysis, and the application of microcomputers and microprocessors to laboratory analysis, have revolutionized our ability to study ecosystems. These innovations have made it possible to perform thousands of analyses where previously only a few could be undertaken by using large amounts of sample and crude reagents. As a result, the cost of analysis has decreased dramatically, bringing this kind of research methodology within the reach of countries and agencies with limited resources. Nevertheless, development projects that include long-term, site-specific research will be more costly and require sustained commitment of funds.

Long-term, site-specific studies require multidisciplinary teams of researchers able to integrate their investigations in an ecological framework. All relevant fields should be represented, including ecology, botany, zoology, microbiology, entomology, pedology, anthropology, economics, conservation biology, agronomy, forestry, and resource management specialties. The research program should specifically include studies of local cultural conditions and resource management practices. The combined study of ecologically important sites and of resource use techniques should serve to protect those sites specifically and the resource base in general.

The number of study sites need not be large. The high costs and limited availability of trained scientists also make it necessary to focus on a few carefully placed sites. Training programs should enable local scientists to acquire technical expertise, thereby compensating for the lack of local professionals (and the small number of foreign professionals in certain disciplines). Local scientists and paraprofessionals should be enabled to take over projects within a reasonable time, depending on local circumstances. So that local scientists can apply their training in the implementation of national conservation programs, it will be essential to strengthen the capacity of their national institutions to support and conduct long-term research, particularly in the areas of ecosystem analysis, meteorology, soil science, conservation biology, and resource management.

Goals and Guidelines

The overall goal of site-specific research, as described in *Research Priorities in Tropical Biology*, is to understand "how natural systems operate in processing and controlling resource flows, and . . . to be able to predict the effect of modification, by management schemes, of the temporal and spatial distribution of ecosystem resources" (NAS, 1980). The report noted that research must combine traditional energetic and watershed approaches to ecosystem studies with a more detailed view of the biology and ecological roles of resident organisms. These remain valid general goals and guidelines for the selection of research sites in developing nations of the tropics and elsewhere. In the intervening years, our awareness of the importance of biological diversity in the functioning of ecosystems has deepened, confirming the need to focus on diversity as a research topic and, hence, as a criterion in the selection of research sites.

The 1980 report also recommended that tropical studies encompass a broader range of primary and modified ecosystems (including cutover forests, naturally reforested clearings, tree plantations, and various agroforestry systems) and devote more attention to the role that specific management strategies had played in reducing biological diversity (NAS, 1980). These goals also remain valid, and should be applied generally to ecosystems in both tropical and non-tropical developing countries. Specifically, research should aim to do the following:

- Understand the basic operations and interactions that characterize primary and secondary ecosystems.
- Obtain information about the adaptive responses of organisms before opportunities for study are lost.
- Identify ecosystems (and components) that are in most urgent need of conservation and preservation.
- Provide an ecologically sound foundation for assessing, managing, and monitoring secondary ecosystems in the tropics and elsewhere.
- Use knowledge of natural regenerative capacities to direct the restoration of degraded lands and improve watershed management.
- Find ways to take greater advantage of the potential productivity of tropical agroecosystems.

Research sites should offer the opportunity to meet these research goals as well as the others described in this report, and should receive a commitment of support for at least a 20-year period. These aims are congruent with those of Project 1 of the U.N. Man and the Biosphere program, and national ecosystem studies should be coordinated with MAB's regional projects and their associated training activities (Unesco, 1974).

Recommended Areas for Site Selection

Research Priorities in Tropical Biology (NAS, 1980) included detailed recommendations for site selection. Again, this discussion remains current, and the findings are summarized here. The rationale behind site recommendations may be found in the 1980 report.

Sites recommended for selection included areas representative of biomes that are highly diverse, in immediate danger of extirpation, located in countries with a history of support for activities of this sort, and logistically convenient. The following locations were recommended:

- Central Brazilian Amazonia;
- La Selva site of the Organization for Tropical Studies in Costa Rica;
- Gunung Mulu National Park in Sarawak, Malaysia;
- Estacion de Biologia at Chamela, Jalisco, Mexico;
- Puerto Rico and Hawaii;

- Barro Colorado Island field station of the Smithsonian Institution;
- Savannas of East Africa; and
- Other grassland areas such as the llanos of western Venezuela and eastern Colombia, the cerrados of Brazil, and the savannas of West Africa.

Research Priorities in Tropical Biology (NAS, 1980) also drew attention to the need to study relatively neglected aquatic systems in the tropics and recommended the following for intensive study:

- Large rivers: the Amazon and Orinoco rivers and their main branches, and the Zaire;
- Smaller rivers: the Musi River in Sumatra, the Purari River in Papua New Guinea, and other rivers similarly threatened with change;
- Lakes: Lakes Valencia and Maracaibo in Venezuela, Lake Malawi in Africa, Lake Titicaca in South America, the volcanic lakes of insular Southeast Asia, closed-basin lakes, lakes that contain major assemblages of endemic species, and lakes subject to change in the immediate future;
- Wetlands: the Sudd along the Nile, the Pantanal of Mato Grosso in Brazil, the Territorio Amapi in Brazil, those of Beni Department in Bolivia, and the Bangweulu Swamp of Zambia;
- Major riverine wetlands: the várzea of the Amazon Basin, the delta and backwaters of the Orinoco, and the backwaters of the Zaire and Xingu; and
- Peat swamps of Southeast Asia.

Since 1980, several important site-specific research initiatives have been implemented in the other developing countries, and should also be considered as potential sites. These include, prominently, Guanacaste National Park in Costa Rica, which was not envisaged a decade ago, and Manu Park Preserve in Peru, to which it was then more difficult to gain access. Abramovitz (1991) provides a comprehensive list of ongoing research activities in developing countries, many of which have the potential to be expanded into long-term ecological research projects (see also McNeely et al., 1990).

In addition, important marine sites need to be considered, especially (from both a biodiversity and a development standpoint) estuaries, mangrove forests, and coral reef ecosystems. Marine systems present special challenges in terms of long-term ecosystem research, and few examples of such projects exist. One such model is the California Cooperative Oceanic Fisheries Investigations (CalCOFI) program, which has monitored the California Current Ecosystem since 1949 (Thorne-Miller and Catena, 1991).

In considering these and other possible sites for selection, the MAB model for pilot projects should be followed if appropriate. Development agencies cannot support this type of activity in all countries, nor is it necessary for them to do so. Such activity is most appropriate in regions where sponsored projects involve major modification of the ecosystem (watershed development projects, for example, including large dams such as the Mahaweli in Sri Lanka, the Manantali in Senegal/Mali, and the Bardheera in Somalia) or where projects establish or provide support to a conservation area or park surrounded by zones of increasingly intense exploitation. Examples of U.S. Agency for International Development (AID) projects that fit this category include the Manu Park Project in Peru, the Cordillera Central Project in Costa Rica, the Masaola Peninsula Project in Madagascar, the Bukit Baka/Bukit Raya Project in Indonesia, and the Korup Preserve in Cameroon. If sites selected for intensive investigation are adjacent to or surrounded by MAB biosphere reserves or other natural areas (as, for example, in the case of the Bukit Baka/Bukit Baya Project), this would allow research to continue far into the future and to focus on the examination and application of principles in conservation biology (see the following recommendation).

Research Priorities

At all long-term research sites, basic information is needed for comparison with other studies. Baseline ecosystem research should include investigations of climatic and geologic influences, water and nutrient cycling, soil chemistry and physics, primary production and ecosystem energetics, biological diversity and species richness, physiological plant ecology, trophic structures, herbivory, higher-order food webs, dynamics of microhabitats and patches, patterns and frequency of site disturbance, human impacts, and ecosystem stability (NAS, 1980; NRC, 1986a).

Research on biological diversity will be only one aspect, albeit fundamental and cross-cutting, of an overall program at long-term research sites. High priority should also be given to topics that are applicable to important management problems, including sustainable agriculture, forestry, and fisheries management. Many sites identified for detailed ecosystem studies are adjacent to or include agricultural areas. Studies of agroecosystems and agroforestry practices in the context of long-term ecological research, conducted in coordination with agricultural scientists, would be both theoretically instructive and of practical use in formulating sustainable production systems for the regions under consideration (NAS, 1980; NRC 1991d).

Given the scope and urgency of the situation, it is necessary to make hard choices among the problems to be studied and the particular areas that should receive attention. These choices should be guided by the potential scientific significance of particular organisms, ecosystems, biological interactions, and indigenous resource management patterns, especially in light of their applicability to human welfare. These choices must also be guided by urgency—the recognition that imminent change might altogether preclude future study of certain organisms, ecosystems, biological interactions, and resource use patterns (NAS, 1980). Specific agendas relevant to the aims of the long-term site-specific research may be found in many documents; see, for example, *Long-Term Studies in Ecology: Approaches and Alternatives* (Likens, 1989), and others noted in the following section.

Conservation Biology Principles and Methods

Research on biological diversity in developing countries should focus on the application and further development of the methodologies and principles of conservation biology

Within the scientific community, conservation biology has emerged as an integrated approach to the questions and concerns raised in this chapter and in other parts of this report. The many basic and applied sciences, academic disciplines, and resource management professions affiliated with conservation biology have worked together to implement more effective means of protecting and managing biological diversity (Soulé, 1986).

The science of conservation biology has special relevance in the context of developing country needs. The implementation of conservation strategies in developing nations, particularly the establishment of biological reserves and parks, presents an opportunity to test the sustainability of conservation concepts and practices. Most of these originated in developed nations of the Temperate Zones, where human population pressures are much lighter than in the tropics, and where ecosystems are generally less diverse. Applying conservation technologies without first testing their effectiveness may yield results similar to those experienced when agricultural technologies developed in the Temperate Zone are transferred without modification to the tropics. Testing and comparing conservation methodologies may enable us to elucidate principles that can be more widely applied.

In a sense, the governments in developing nations are already performing large experiments with biological diversity, often with funding provided by international development and lending agencies. The building of a road or dam, the installation of an irrigation project, the implementation of a resettlement program, the expansion of agriculture, or the intensification of resource extraction—all are experiments that affect the status of biological diversity. They are prevented from being research experiments in the true sense by the fact that data on the effects of the experimental treatments are not collected and analyzed systematically, and no controls are maintained for purposes of comparison. In many cases, research on biological diversity in the strictest sense could be advanced simply by requiring that development agencies conduct their experiments correctly. This problem, of course, also exists in developed nations: the type of long-range monitoring that would allow information about these activities to be applied to subsequent development projects is simply not supported or implemented adequately anywhere.

Given the number of species presently threatened with extinction due to habitat destruction, the lack of understanding of their basic biology, and the expense (in terms of money, energy, and constant human attention) involved in maintaining species outside their habitats, in situ conservation strategies are the most practical, cost effective, and dependable for the vast majority of organisms, particularly in the tropics. Establishing and maintaining protected natural areas are essential parts of the task. These must be planned and managed so as to embrace a maximal range of habitats and to create as effective a system as possible.

As a rule, on-site conservation works only where there is local commitment to it. This is especially important in developing nations, and effective conservation methodologies must be tailored to reflect this need. Development agencies, recognizing the importance of conserved areas as a source of future development options, have become involved in helping national governments and private voluntary organizations to establish and manage conservation areas. They have increased support for local involvement in the training of conservation and wildland managers, the formation of conservation strategies, the search for economic uses of conserved areas, and the development of substitutes for products now obtained from endangered species and habitats.

Providing an improved scientific basis for the conservation of species and habitats requires investigations into all aspects of their biology. For all species—but especially for those that are rare, threatened, or declining—research must illuminate their taxonomic status, genetic variation, life histories, population dynamics and ecology, distribution, habitat needs, and the effect of human activities on their circumstances. The impact of introduced species on biological diversity in many ecosystems, in particular island ecosystems, has been significant. In such cases, research should determine the role of these species and, if necessary, on possible ameliorative measures.

At the ecosystem level, research should focus on securing an effective system of natural areas and reserves. Attention must be given to basic concepts in the design and establishment of reserves—the identification of "hot spots," the size and configuration of protected areas, their coordination and linkage, the dynamics of natural disturbance regimes, and the effectiveness of buffer zones—and to the integration of these reserves with local communities and land use patterns. Research must synthesize these and other factors, taking into account the many scales—local, landscape, ecosystem, national, regional, and even global—at which conservation must operate to be successful.

Although ex situ methods are able to preserve genetic diversity only to a limited extent, these methods require increased research and emphasis. Research programs should include provisions to establish or strengthen systems of botanical gardens, zoos, stock culture centers, and captive propagation programs for the preservation of selected organisms that cannot survive in the wild. Ex situ methods are also highly important as part of an overall strategy to conserve wild and domesticated plant and animal genetic resources. Greater support is needed for the improvement of seed banks, crop collections, and other types of genetic reservoirs (NSB, 1989; NRC, 1991b, 1991c).

Research Priorities

Research Priorities for Conservation Biology (Soulé and Kohm, 1989) is a complete guide to research needs in conservation biology, and is widely applicable to the challenges of conservation in developing countries. Other recommended documents are *Research Priorities in Tropical Biology* (NAS, 1980); *Ecological Knowledge and Environmental Problem Solving: Concepts and Case Studies* (NRC, 1986a); *The Sustainable Biosphere Initiative: An Ecological Research Agenda* (ESA, 1991); *Funding Priorities for Research Towards Effective Sustainable Management of Biodiversity Resources in Tropical Asia* (Ashton, 1989); and *From Genes to Ecosystems: A Research Agenda for Biodiversity* (Solbrig, 1991). Development agencies can provide an important service by making these documents more readily available to scientists, resource managers, agencies, and other institutions in developing countries.

Sustainable Use of Biological Resources

Research should be conducted on strategies for the sustainable use of biological diversity and for returning something of the value of biodiversity to developing countries.

Sustainable use implies that current human needs can be met without degrading the resource base for future generations. Although many strategies for accomplishing this have been advanced, few have undergone scientific scrutiny. Substantive research results are needed to guide policymakers in choosing among these (NRC, 1991d; ESA, 1991).

Much of the emphasis in conservation has involved securing and managing protected areas of special concern, as discussed in the previous section. As critical as this is, conservation of biological diversity in developing (as well as developed) countries cannot be achieved through the establishment of protected areas alone. Conservation and management measures must apply to surrounding land uses as well. Strategies for sustainable use must complement protection efforts, and countries and localities that use biological resources on a sustainable basis need to benefit financially for their efforts.

The establishment of extractive reserves in tropical forests is one strategy that has recently attracted much attention. Such reserves are managed to provide access for local people to extract or harvest products in a nondestructive fashion. Recent studies have demonstrated dramatically the potential value of tropical forests for extraction activities (Peters et al., 1989). Extractive reserves, however, are limited in the extent to which they can provide an adequate livelihood for forest dwellers, since beyond a certain limit of market saturation extraction will drive the price and income down, and encourage substitutes or more intensive (and potentially harmful) production methods. Research is needed on the economics of this process to identify both its potential and its limits.

A second strategy with great potential is the so-called debt-for-nature arrangement, in which portions of the international debts owed by developing countries are purchased by intermediary groups—in practice, private conservation groups have served in this role—and exchanged for other equity, usually funds in the currency of the debtor government. Under these arrangements, debtor governments agree to devote land and funds to in-country conservation activities. In Ecuador, for example, the World Wide Fund for Nature helped to negotiate a debt purchase, the funds from which will be used to support park management and conservation education. Debt exchange is still a new tool for conservation, and will not by itself relieve either the debt crisis or the pressures on biodiversity. Nonetheless, it holds great potential, and biologists should work together with economists and policy experts to adapt and refine it.

In addition to such specific strategies, the conservation of biodiversity must also be pursued through the accelerated development and implementation of sustainable agroecosystems (NRC, 1991d). Biological diversity cannot be conserved effectively unless human demographic pressures are addressed and alleviated—that is, unless people have enough food, income, and social stability to prevent overexploitation of resources and continual movement into less exploited areas (especially those of high biological richness). To meet these needs, all productive land uses must be modified and better coordinated at a landscape level so as to safeguard those biotically diverse features they retain. A variety of agroforestry and agropastoral systems, home gardens, modified forests, and other sustainable land use systems—many of them based on traditional management techniques—can be applied toward this end, and special emphasis should be given to their use on degraded and abandoned lands (see following section).

Research Priorities

- Determine the capacity of minimally disturbed forest to provide useful and valuable products on a sustained basis;
- Develop the means to secure income for local people and communities through local processing, the protection of intellectual property rights, commercial agreements and investments, and incentives for the adoption of sustainable land uses.
- Determine how the juxtaposition of intensively managed systems (agriculture and plantation forestry), lower intensity land use systems (e.g., agroforestry and home gardens), and neighboring forest reserves affects productivity and biological diversity within landscapes;
- Determine the influence of natural areas on managed agroecosystems (through, for example, pollinators, pests, predators, parasites, and antagonists; nutrient-pumping associations; and moisture conservation and microclimatic effects); and
- Determine how the productivity of degraded and abandoned lands can be enhanced and sustained so that pressure on remaining natural systems is reduced.

Restoring and Using Degraded Lands

Increased support should be given to research on the restoration and utilization of degraded lands and ecosystems in developing countries.

Heavy resource exploitation—deforestation, overgrazing, soil erosion, nutrient depletion, and salinization; pollution; overharvesting and mismanagement of fisheries; surface mining; disruption of hydrological systems; and indiscriminate wetland drainage and filling—has left landscapes and ecosystems in both developing and industrialized nations in a state of biological and social impoverishment. Efforts to restore these areas of acute degradation, as well as partially damaged ecosystems, are essential if the cycle of environmental decline, nonsustainable land use, and socioeconomic instability is to be broken.

Until now, efforts to conserve biological diversity have concentrated primarily on inventory and classification, in situ and ex situ preservation, the exploration of indigenous knowledge, and the development of techniques and policies in support of sustainable use. These endeavors, as evidenced by the recommendations in this report, must continue and must expand. At the same time, the restoration of degraded lands and ecosystems should assume a more prominent position in the spectrum of conservation activities. Not only must the loss of biological diversity be arrested, but the damage must be repaired to the extent feasible (Cairns, 1988). Restoration allows us to see in degraded lands not just a past failure to conserve, but a future opportunity to conserve—to rebuild biological diversity for its own sake and for the benefit of the people whose livelihoods depend on it.

The application of restoration techniques provides many benefits. Restoration requires source pools and baseline information on ecosystem functions; this, in turn, is an additional and often overlooked rationale for the protection and long-term study of natural areas and ecological reserves. By recreating lost habitat, supplementing that which exists in protected areas, and serving as gene pool reservoirs, restoration efforts can provide directly for the conservation of biological diversity. Restored forests, pastures, range, arable soils, wetlands, and aquatic systems also provide critical environmental services by protecting soils, moderating hydrological processes, sequestering carbon dioxide (the principal greenhouse gas), catching sediments and pollutants, and serving as buffer zones (Jordan, 1988).

Restoration also yields important social and economic benefits. Returning systems to productivity will enhance the economic value of these systems as extractive reserves and for sustainable agriculture, agroforestry, forestry, fisheries, and livestock production while helping to relieve the pressure to exploit intact land. In addition, active programs of restoration offer opportunities for social services through the education, training, and employment of personnel necessary to implement them. These benefits will often accrue where they are most needed: in areas where resource degradation has not only depleted biological diversity but led to ingrained poverty and destabilized rural communities.

Research on restoration is especially important in the humid tropics, where degraded lands currently support few people and where restoration directly involves the preservation and enhancement of biological diversity and the encouragement of sustainable uses. It is estimated that more than one billion hectares of degraded lands have accumulated in tropical countries over the last several decades, and that about 700 million hectares are in need of reforestation (Grainger, 1988). These lands can potentially be used for agriculture, agroforestry, plantation forestry, sustainable livestock production, human settlements, and other managed land uses. At present, however, most of these lands lie degraded, regenerating as conditions permit but usually producing little benefit to man.

In addition to reducing the pressures on the remaining tropical forest, bringing degraded lands back to production would very likely supply all the food, fiber, clothing, shelter, and fuel needs of all the people in the tropics who presently must resort to deforestation. Reforestation can also play an important role in slowing the buildup of atmospheric carbon dioxide, a contribution that would be further augmented if new forests were managed to provide renewable energy sources as substitutes for fossil fuels (Houghton et al., 1990). Given the current rate of population increase, these areas must eventually be restored to some level of sustainable use. It makes sense, therefore, to concentrate efforts on tropical forest restoration as a matter of urgency.

Restoration research, though practiced on a small scale for a number of years in some developed nations, is a relatively new field of investigation in developing countries (Jordan et al., 1987; NRC, 1989). Currently, only a limited theoretical foundation can be applied in site restoration, and there are very few cases in which these theories have actually been tested (e.g., Janzen, 1988; Uhl, 1988; Uhl et al., 1990; Nepstad et al., 1991). As critical as it is to implement restoration in tropical forest regions, these efforts should be extended to other areas and systems in developing countries, including wetlands, riparian zones, hill lands outside the tropics, coral reefs, rangelands and grasslands, areas affected by salinization, and areas affected by mines and other industrial developments.

Development agencies must play a larger role in encouraging these efforts and applying restoration techniques more widely. As knowledge builds with experience, it will become possible to derive generalized principles of restoration ecology and management, including identification of the constraints on and opportunities for restoration under different ecological conditions and circumstances; the purposes and benefits of different restoration "regimes"; the impact of site size on potential restoration strategies; and the coordination of restoration with other conservation activities.

Unfortunately, there is ample space in which to experiment. At this time, it is important to encourage experimentation by supporting research and demonstration projects across a range of ecosystems in a variety of social and cultural settings. As of now, there are few institutions or professionals with the necessary expertise in restoration.

In the long run, however, the training of scientists and technicians in developing countries must be a significant component of restoration programs.

Research Priorities

• Study the impacts of different prior landuses on restoration potential, identifying the principal factors that affect restoration in different systems.
• Establish thresholds beyond which ecosystem recovery from anthropogenic disturbance will not occur.
• Study and compare natural and anthropogenic disturbance regimes in ecosystems, and succession and recovery processes (both natural and manipulated) in degraded sites.
• Compare the rate of ecosystem recovery when different mixes of species are used for site restoration (e.g., trees; trees and shrubs; or trees, shrubs, and herbs).
• Study and compare the impacts of introduced exotic species under controlled conditions to prevent their escape before the potentially detrimental effects of their use in restoration are understood.
• Establish the site requirements for the reintroduction of specific forest species with high forestry and fuelwood potential.
• Compare the physical and biological properties of disturbed areas where trees are reestablishing and where they have not reestablished; identify causal factors (in particular, the influence of soil biota) and evaluate different species for their ability to revegetate disturbed areas;
• Identify the social and economic constraints on, and opportunities for, restoration work in different systems; compare the environmental, economic, and social implications of restoration through different agroforestry and annual cropping systems.

INFORMATION NEEDS

To be most useful, scientific information on the extent, status, value, use, and preservation of biological diversity must be coordinated, accessible, and applicable. This is especially important in developing nations, where inadequate infrastructure, information technologies, and networks can be primary constraints on research and effective conservation activities. Because of their extensive experience and institutional structures, international development agencies can play a vital role in overcoming these constraints. As they do, it is important to remember that the channels of communication for information on biological diversity must remain open. Information must both be made available to and draw on the work of scientists, resource agencies, national institutions, and nongovernmental organizations in developing countries.

In the decade since the publication of the NAS (1980) report *Research Priorities in Tropical Biology*, the rapid evolution and increasing availability of information technologies, in particular personal computers and geographical information systems, have revolutionized our ability to organize and analyze information. This has significant implications for the conduct and application of biodiversity-related research. Information networks, in particular, not only allow in-country researchers to take advantage of the work of other scientists but give them a greater sense of purpose and a broader understanding of the context in which they are working. This knowledge is especially important to scientists working in the same region (such as the Amazon basin), in similar systems from different regions (such as tropical rain forests), and on elements that move between regions (such as neotropical migrants).

Scientific information need not be disseminated through the highest technology to be of significant use. A modest field guide or parataxonomist training manual, properly designed and distributed, may be more effective in terms of real needs and real results than a new computer program or satellite image. Determining and coordinating local, national, and regional information needs represent major challenges for development agencies. Several of the most significant information needs—computer inventories, identification and classification programs, remote sensing capability—have been mentioned in the context of previous recommendations.

To enhance the availability and application of scientific information for the purposes of managing and conserving biological diversity, the following actions are needed.

Computer Data Bases and Inventories

Resources should be devoted to the development of computer data bases, inventories, and information networks for the collection and collation of information. Support should be given to the improvement of interinstitutional coordination, system design, and operational administration through the establishment of national biological institutes or equivalent centers.

As conservation faces greater competition for resources, the need for coordination and shared information to prevent duplication of efforts becomes paramount. To select and design new reserves, appropriately manage and monitor existing reserves, take advantage of opportunities for sustainable land use and restoration, and coordinate ex situ conservation efforts, researchers and administrators must have access to information on the classification, distribution, characteristics, status, and ecological relationships of species. Much of this information, if it exists, is scattered and difficult to obtain. The development of computer data bases and inventories would be a major factor in overcoming this constraint. National biological institutes can provide a central location for these data bases, inventories, and information networks and promote the interinstitutional coordination necessary to their success.

Many current computer programs specialize in the long-term management of information necessary for the conservation of biodiversity. In the United States, these include the Heritage and Conservation Data Centers (CDCs) of the Nature Conservancy; the data bases of botanical gardens, arboreta, museums, herbaria, aquaria, and zoos; the breeding bird and waterfowl surveys of the U.S. Fish and Wildlife Service; the Christmas Bird Counts of the National Audubon Society; and the lepidoptera surveys conducted by the Xerces Society. Notable among examples in other countries are the data bases of INBio in Costa Rica.

In 1974 the Nature Conservancy initiated the first of what are now known as State Natural Heritage Inventories. This effort has been expanded to include all 50 states in the U.S., 2 provinces of Canada, and 13 Latin American and Caribbean countries. The Nature Conservancy's CDCs are continually updated inventories of the most significant biological and ecological features of the country or region in which they are located. These computerized centers offer a readily accessible source of information on biological diversity that can be used in conservation and development planning. All 70 CDCs now operating in the Western Hemisphere employ the same methodology and computer hardware. The CDCs in Latin America are staffed and operated by local scientists and conservationists.

A quite different example is Tropicos, the botanical data base of the Missouri Botanical Garden, which serves as a tool in systematic research and in the production and revision of flora. It includes programs for managing herbaria, producing herbarium specimen labels, maintaining horticultural information on living specimens, and managing botanical libraries. The data base currently contains about 400,000 names of taxa. Entries include information concerning synonyms, literature citations, description, and distribution (at the country and two additional subunit levels). The system also has the capacity to generate plant descriptions for floras, character lists, chromosomal analyses, and information on the taxonomic status of specimens.

Jenkins (1988) lists the following principal uses of these kinds of data:

- Facilitating continuing inventory by organizing data well enough to determine what is and is not known;
- Setting and revising conservation priorities through an ever-improving picture of the relative endangerment and status of species, habitats, etc.;
- Selection and design of reserves through the identification of areas containing critical species or habitats and an understanding of their ranges and needs;
- Facilitating more efficient and sophisticated use of land protection methods by conservation administrators;
- Monitoring and managing biological elements—a species, community type, or other feature of interest—by enabling users to make rapid field assessments of their status and ecological response to management options;
- Site management;
- Providing information about sensitive sites and project design requirements to developers and development agencies in the planning process;
- Providing real data for environmental impact analysis and review;
- Providing access to additional information by including references to original sources, published and unpublished documents, professional sources, museums, files, data bases, and maps;
- Providing data for extrapolation in predictive modeling; and
- Providing field localities, biogeographic information, and other baseline data necessary for research concerning conservation principles.

Given the myriad applications of a coordinated, well-designed, and well-maintained data base network, this will clearly be an important tool for developing countries. Fortunately, the examples cited above, as well as most others, can be run on personal computers that do not require large investments in software or hardware. Much of the required software can be obtained at cost or free of charge.

Remote Sensing and Geographic Information Systems

Additional research and technical development are needed to advance the utility of remotely sensed data for ecosystem monitoring in developing countries.

The data from remote sensing techniques, coupled with the data management capacity of geographic information systems (GIS), offer unprecedented opportunities to assess and monitor ecosystem processes. Even regions that are experiencing rapid change, such as tropical environments, can be closely surveyed through means not available a decade ago. Computerized geographical information systems have, in this same period, simplified the process (and decreased the expense) of adding new data and adjusting analysis.

No one source of remotely sensed information is likely to supply all of the data to address biodiversity research needs in developing countries. Coarse spatial resolution sensors with high rates of data acquisition (e.g., the Advanced Very High Resolution Radiometer of the National Oceanic and Atmospheric Administration) are needed to accommodate the vast land areas studied in tropical surveys. High-resolution sensors—for example, the Landsat MSS (Multispectral Scanner) and TM (Thematic Mapper), the Systeme Probatoire d'Observation de la Terre (SPOT), aircraft scanners, and mapping cameras—are needed to record spectral and spatial information to link intensive field-level ecological studies to forest community and biome-level assessments. In regions with frequent cloud cover, sensors that operate in the visible and near infrared have limited utility. In these areas, active microwave sensors can provide information about the land surface and forest canopy that would otherwise be unobtainable (Sader et al., 1990).

The benefits of geographical information system technology go far beyond its ability to maintain information in a geographically referenced format. Information on soils, topography, climate, distribution of organisms, land use, and protected areas can both clarify and augment the measurements provided by remotely sensed data (Green, 1981).

Gap analysis is one important application of remotely sensed data (Burley, 1988; Scott et al., 1991a, 1991b). Laws, policies, and, to a great extent, public opinion tend to focus our financial and intellectual resources on individual species. Gap analysis identifies gaps in the network of protected areas and compares these against the background of the distributions of ecosystems, vegetation types, and plant and animal taxa. Although gap analysis is still being developed, it holds great promise. Gap analysis can reveal priorities for conservation in a more systematic and quantified manner than previous methods, and can pave the way for the protection and management of sensitive areas. By adopting a broadly based ecosystem approach, it seeks to prevent species and communities from becoming endangered, allowing scarce human and financial resources to be applied more effectively.

Other applications of remotely sensed data are already in use and producing much-needed information on the status of resources in tropical regions. Remote sensing was used, for example, to estimate available habitat for migrating birds in the Yucatan of southern Mexico (Green et al., 1987). In Thailand, Vibulsresth (1986) was able to differentiate "disturbed" from "undisturbed" dry dipterocarp forests.

Perhaps the most notable use of remote sensing data was the publication in *The New York Times* of images of the burning forests of Rondonia in Brazil (Matson and Holben, 1987).

Difficulties involved in developing remote sensing and geographical information system capabilities include a lack of continuity in the data, the cost of data, and the lack of equipment and training opportunities (Grainger, 1984). Coordination of these research activities (again, within national biological institutes or centers) is also needed. If these problems can be overcome, programs at the regional and global level can proceed by using data from sensors already in orbit.

Strengthening Scientific Networks

Development agencies should use their financial and institutional resources to establish and encourage networks that foster communication among scientists working with biological diversity in developing countries.

The effectiveness of all scientists depends in large part on their access to professional colleagues and to information in their field. Those who study biological diversity in developing nations face special difficulties. Traditional sources of scientific information—libraries, museums, universities—often lack the resources to maintain up-to-date collections and to disseminate the findings of their own researchers. The cost and inconvenience of travel to scientific meetings and conferences can be prohibitive; modern communication technologies are often unavailable. As the need for scientific information on biological diversity grows, and as the volume and quality of information increase, scientific networks must keep pace. These networks should serve to improve communication among scientists in developing countries, between scientists in different countries, and between scientists in the developing and the developed world.

Support for scientific networks begins at the field research level, with increased financial support for operations and data analysis. The development of methods for reporting data and managing information, particularly computerized inventory data, has been discussed above. Scientific networks will play a leading role in refining these methods, coordinating research efforts, and improving the channels of communication from the field to the international level. Development agencies can best support this work by backing existing networks, such as the Latin American Plant Sciences Network (see sidebar) and the Association pour l'Etude Taxonomique de la Flore d'Afrique Tropicale, and by promoting the establishment of similar networks in regions where they currently do not exist.

A number of steps that development agencies can take to improve

The Latin American Plant Sciences Network

Red Latinoamericana de Botánica (the Latin American Plant Sciences Network), or RLP, is a consortium of six graduate training centers located in Mexico, Costa Rica, Venezuela, Brazil, Chile, and Argentina. In these centers, Latin American academic institutions collaborate to provide graduate level training in the plant sciences to students from throughout in the region. The centers also organize binational research projects, regional graduate courses, and scientific meetings.

The primary aim of the network is to promote development of the plant sciences in an indigenous context in the countries of Latin America. It seeks, in particular, to strengthen the capacity of these countries to conduct basic research and training in biodiversity, conservation, and sustainable agriculture. In addition to providing educational opportunities for plant scientists, RLP supports the development of new centers of botanical excellence throughout Latin America, increased interaction among Latin American scientists, and the promotion of regional self-sufficiency through strengthened international relations.

The network, which was established in 1988, also works with non-Latin American institutions in achieving these objectives, and has received support for its activities from the U.S. Agency for International Development and a number of private foundations (RLB, 1991).

communication among scientists in developing countries would directly promote the formation and strengthening of networks:

- Improve access to bibliographic resources and other data bases by providing scientific and educational institutions with funds for journal subscriptions and book purchases.
- Support the publication of findings in international journals and local publications, especially those in vernacular languages (a considerable amount of data on the floras of many countries has gone unpublished for lack of funds).
- Require that proposals for agency-sponsored research in developing countries include funds in their budgets for the publication of results in a form accessible to scientists in the countries themselves.

- Support the publication of newsletters.
- Finance the compilation of a worldwide directory of individuals working in the area of local knowledge systems, and support the preparation and publication of annotated bibliographies on selected topics related to local knowledge.
- Sponsor local, national, and regional workshops and conferences on biological diversity, and provide increased funding for scientists to attend international conferences.

HUMAN RESOURCES

All conservation work in developing countries, but especially basic research on biological diversity, is hindered by a lack of trained personnel. The recommendations offered below have been mentioned earlier in this report, but are reiterated here because of their fundamental importance. In some countries it may simply be impossible to carry out specific projects suggested in this report because few local scientists or resource managers have the necessary training and experience. Taxonomists and conservation-oriented ecologists and biologists, scarce to begin with, are overwhelmingly concentrated in industrialized nations. At the same time, career conservationists in developing countries must often contend with a lack of support from their own governments. Thus, the implementation of conservation programs in developing countries is often contingent on the availability of foreign expertise and the continued willingness of host countries to have their conservation programs dependent on foreign nationals.

This dangerous situation must be remedied if the conservation of biological diversity is to become a continuing, ingrained activity in developing countries. In turn, both the fostering of a strong corps of local trained conservation personnel and the strengthening of institutions that guide, support, and coordinate their work, are necessary. In this context, the role of international networks is particularly important, including, for example, the Latin American Plant Sciences Network; the programs of the International Council of Scientific Unions (ICSU); the Third World Academy of Sciences; the African Academy of Sciences; and the Unesco-supported African BioSciences Network.

The urgent need for manpower to formulate and conduct research and applied conservation activities prohibits the lengthy process traditionally employed to train ecologists, taxonomists, or resource managers. There is simply not enough time to produce enough people by these methods. The situation demands new types of paraprofessionals and new ways of producing them. There is also a desperate need to strengthen the capabilities of in-country agencies and institutions responsible for the conservation and management of natural resources. Development agencies, charged with institution and manpower development, have the unique experience, capabilities, and resources to address these problems.

To strengthen the human resources necessary to survey, research, monitor, and manage biological diversity in developing nations, the following actions are needed.

Developing Taxonomic Expertise

International development agencies should sponsor and support the development of taxonomic expertise, both paraprofessional and professional, as an increasingly important part of their conservation programs.

Many of the recommendations outlined presume the existence of the taxonomic expertise to carry them out. Yet the cadre of trained taxonomists necessary to perform this work simply does not exist. To describe, inventory, classify, monitor, and manage biological diversity, such expertise must be cultivated. It is the foundation on which the study and conservation of biological diversity are built.

The situation is not new, and the call for a response has been heard before. The report *Research Priorities in Tropical Biology* (NAS, 1980) recommended that "high priority . . . be set on training and support for much larger numbers of systematists oriented toward tropical organisms. At least a five-fold increase in the number of systematists is necessary to deal with a significant proportion of the estimated diversity while it is still available for study. Governments would be well advised to allocate resources in an effort to achieve this objective." Since the 1980 report, little progress has been made in meeting this fundamental need. Another decade of neglect cannot be allowed to elapse.

Taxonomic expertise for certain groups of organisms is almost nonexistent. For example, termites and ants constitute approximately 30 percent of the world's animal biomass, and are of enormous ecological significance. Yet there are less than a few dozen people worldwide who are able to classify, or even competently sort, specimens. There are not even five people in the world who can identify, sort, or characterize free-living nematodes. Some 12,000 species of nematodes have been described, including all that are animal parasites and cause diseases in domesticated plants. It is estimated that a million species of free-living nematodes may exist worldwide, but because there are no systematists working with them, no organized evaluation of the diversity in this group can be made. The story is similar for mites. Again, it has been roughly estimated that a million species exist, yet only 30,000 have been described. Although it may not be possible to support specialists in these groups in all developing countries, it is imperative that the expertise to at least sample these species be encouraged.

Many factors have contributed to the paucity of trained personnel, including a shortage of research and teaching positions for systematists, the lack of training grants, and competition from other areas of biology (NSB, 1989). Academic departments throughout the world have been trading organismal biologists for molecular biologists, with the result that fewer undergraduates are exposed to taxonomy and, hence, given the opportunity to pursue these fields in graduate schools. The situation must be changed through the creation of more positions for taxonomists at all levels and of programs to inform students interested in taxonomy of the opportunities that exist. In addition, support for research should be made available to students and faculty alike.

These problems are especially prevalent in developing countries, where the need is most evident. The situation is exacerbated by poverty, scarce funds, inadequate institutional support, and a general lack of trained native scientists. Although developing countries contain 77 percent of the world's people and 80 percent of the world's biodiversity, they have no more than 15 percent of the world's wealth, and only 6 percent of the world's scientists and engineers live and work in them. The industrial countries and development agencies can do much to build a base of taxonomic expertise by providing more amply in their assistance programs for strengthening the institutions in which taxonomic work is based (see following recommendation).

The positive side of this situation is that great rewards can be obtained by employing nontraditionally trained people in this work. The notable example of this, again, is INBio in Costa Rica. It should be noted that in augmenting the taxonomic proficiency of personnel in developing countries, many other areas of scientific research and application will be served. Systematists are indispensable for progress in all fields of basic and applied biology, including ecology, fisheries biology, range management, forestry, agriculture, horticulture, and the veterinary and medical sciences (NSB, 1989).

To promote awareness of the basic importance of taxonomic work internally, international development agencies should develop and conduct crash courses in taxonomy and conservation, using approaches similar to those employed in plant breeding programs. Agencies can also provide an important service by designing and testing data management systems for ease of use by conservation paraprofessionals and for transferability across cultures. In some cases, it may be possible to meet these needs by strengthening existing training centers, such as Mweka in Tanzania, Garoua in Camaroon, Dehra Dun in India,

Bariloche in Argentina, and the Centro Agronómica Tropical de Investigación y Enseñanza (CATIE) in Costa Rica (McNeely, 1989).

Strengthening Local Institutions

Because the fate of biological diversity in developing countries depends ultimately upon the sense of stewardship, the scientific capacities, and the administrative structures within these countries, it is highly important that development agencies invest in strengthening local institutions.

Only native institutions are capable of imparting the understanding of biological diversity to the general public and the proficiency among professionals that will result in effective conservation. It is especially important that development agencies support nongovernmental organizations (NGOs), educational institutions, museums, and libraries in developing countries, and foster effective operation of the government agencies legally charged with managing resources. Unless this local capacity grows, effective conservation will continue to rely too much on the concerns, expertise, and changing priorities of foreign institutions.

The support of museums and libraries in developing countries is crucial. Museums often contain irreplaceable samples of the biota of their countries. In general, unfortunately, they are poorly supported. Whether they exist in universities or government ministries, or as public institutions, museums should receive special assistance from development agencies as baseline institutions for the collection and classification of organisms. Botanical gardens, arboreta, herbaria, aquaria, seed banks, and zoos, although generally uncommon in the developing world, are to be encouraged as essential for the documentation of local biological knowledge, and as locations for ex situ conservation efforts. Specifically, development agencies should assist nations in determining which germ plasm should be conserved ex situ, and which national institutions should be charged with the maintenance of different collections. Personnel should be trained in the latest curatorial techniques.

The public awareness and educational activities of these institutions should also be improved and encouraged to involve local populations more actively—thereby improving the possibilities for recurrent cost recovery. Development agencies should provide seed funding for this kind of activity, with the prospect of phasing out funds as local support is generated from government, voluntary activities, and donations.

It is especially important for development agencies to support those government agencies charged with the protection and management of natural resources (and hence biodiversity) both in the field and in ex situ collections (e.g., ministries of agriculture, forestry, fisheries, energy, and parks). Only through strengthening these legitimate institutions can the preservation and sustainable management of biological resources be ingrained in the society.

The diversity and complexity of ecological, political, social, and economic conditions in developing countries has led to the burgeoning of locally based nongovernmental organizations that serve as important conduits for the flow of information to, from, and among local people and communities. Some of these organizations focus on conservation, but many others with an interest and a stake in land use issues lack the experience, resources, and personnel to follow up on their concerns. National and international development agencies need to support the involvement of NGOs as intermediaries between government agencies, universities, and local communities in support of the methods and goals of biodiversity conservation. Such investments can have profound consequences. In the long term, providing funding and political support for NGOs will be more effective in shaping environmentally and socially acceptable land use policies, based on local needs and priorities, than the dictation of policy by foreign and international governments and institutions.

Expanding Cooperative Research Programs

New and existing programs of international cooperative research should undertake research on biological diversity as a fundamental part of their mission, and should be given the financial and administrative support to do so.

International cooperative research programs that are currently devoted to crop agriculture, forestry, aquaculture, livestock management, and other resource uses in developing countries should give greater attention to biodiversity within their research and development programs. Biodiversity and its relationship to sustainable land use are so central to attainment of development goals that they should be fundamental considerations in carrying out all research programs involving natural resource management. In particular, these programs need to involve more systematists and other biologists to perform basic research on biodiversity in developing countries (NRC, 1991d).

In the past, progress toward sustainable resource management has been hindered by policies and technologies based on discipline-specific research. In the future, land use and resource management must overcome these boundaries, and interdisciplinary research must provide the knowledge to do so (Soulé and Kohm, 1989). The conservation of biodiversity should not be pursued as an isolated area of research, but integrated into the activities of all research institutions and programs that affect land use in developing countries.

Research conducted under the auspices of the International Council of Scientific Unions and Unesco's Man and the Biosphere Program provide important models for the integration of biodiversity studies in a multidisciplinary research framework. In addition, several new cooperative research and training programs have begun to incorporate this approach. For example, within the U.S. Agency for International Development, the Sustainable Agriculture and Natural Resources Management (SANREM) Cooperative Research Support Program and the Program on Scientific and Technical Cooperation (PSTC), a competitive grants program designed to fund innovative research projects, both involve significant biodiversity research components (NRC, 1991d). The Global Environmental Facility (GEF) of the World Bank also promises to give greater attention to research on biodiversity.

While these efforts are to be commended, the general level of financial and administrative support within international research organizations is still far too meager, given the magnitude of the problem. Put more positively, great opportunities exist for constructive and mutually beneficial cooperation between scientists working on the conservation of biological diversity and scientists in other fields of land use, resource management, and rural development. Development agencies should encourage this cooperation—not just as a new aspect of research, but as a new and increasingly necessary way to perform research.

3

Biodiversity Research: The Socioeconomic Context

Resource depletion, which usually involves habitat conversion and the attendant loss of biological diversity, is often justified as the only way for developing nations, faced with growing populations, huge foreign debts, nascent industrial capacity, and a largely uneducated agrarian populace, to develop. This was the course followed by so many industrialized nations, and it has become the standard development pattern worldwide.

This development model, however, is undergoing extensive scrutiny. As more economists have begun to work with, and incorporate the understanding of, environmental scientists, they have raised fundamental questions about the assumptions and prescriptions of conventional economics that underlie this model. (See, for example, Daly and Cobb, 1990; Pearce and Turner, 1990; Costanza, 1991. The dialogue between economists and ecologists has also been advanced through recent establishment of the International Society of Ecological Economics and publication of its journal *Ecological Economics*.) This reexamination is growing increasingly sophisticated in its analysis, but it is based on the simple recognition that the scale and the speed of anthropogenic alteration and depletion of soil, water, atmospheric, and biological resources have disastrous repercussions.

These consequences have, in fact, already overtaken many countries. Areas that have lost biological diversity may become seriously degraded, for example, the salinized soils of northeastern Thailand and the Indus valley. Other areas may be intrinsically less productive, as in the extreme case of the *Imperata cylindrica* grasslands that have spread rapidly over millions of hectares of disturbed and deforested lands in the tropics, rendering them all but useless from an economic standpoint. Alternatively, areas may remain productive but vulnerable to biological disaster. In Asia, leucaena (ipil-ipil), one of the most widely planted and productive multipurpose trees in use today, is now being devastated by a psyllid insect pest. Similarly, the monocultures of cassava in West and Central Africa were ravaged when the cassava mealybug was inadvertently introduced from South America, the original home of cassava. The mealybug was brought under control only through the introduction of a natural parasitic wasp—a move that may have been worth billions of dollars and illustrates the importance of understanding the role of biodiversity in agroecosystem management (Norgaard, 1988; Neuenschwander et al., 1990).

As the resource depletion model of development has been adopted worldwide, the loss of biological diversity has accelerated. Ultimate socioeconomic reasons for this loss include the failure to consider environmental externalities in benefit-cost analyses; misguided government policies; land tenure systems that promote resource depletion; inadequate institutional infrastructure; ineffective communication among local, national, and international institutions; and inequitable distribution of political power and its attendant corruption. Highly destructive global economic forces, including the accumulated $1.2 trillion international debt and, since 1984, the flow of cash away from poor developing nations to wealthy industrialized ones, have reinforced these factors. It is fundamental, therefore, that the development community and the developing countries themselves evaluate the economic factors involved in the use or destruction of biological diversity.

ECONOMIC RESEARCH AND THE CONSERVATION OF BIODIVERSITY

Economic theory provides an analytical method for diagnosing when inefficient use of environmental resources is likely to occur. It holds that resources will be allocated efficiently when prices reflect true resource scarcity, when there exists a right of ownership to resources so that free trading of resources is possible, and when consumers have access to information about the total amount of a resource available. Economists will continue to argue if and when these assumptions are met, but the expanding threats to biological diversity and other natural resources have raised fundamental concerns about the limits of this method as practiced.

First, natural resources provide nonmarketed goods and services as well as commodities. Usually, however, only commodities are openly traded, and therefore priced, by markets. For example, harvested wood may be priced in the marketplace, but trees also provide medicine and other minor products for local peoples, and control soil erosion and flooding regimes. Thus, trees simultaneously provide commodities that may be traded on the world market (logs), goods available primarily through the local market (medicines), and services that are not traded (erosion and flood control).

Second, incentives, tax provisions, credits, subsidies, and other economic policies distort commodity prices and encourage massive environmental transformation. Many developing nations have institutionalized short-term profit taking through resource depletion via these and other economic policy instruments. Tax holidays, inadequate rent taxation, low stumpage charges, and no-interest loans for forest clearing to establish cattle ranches are examples of policies that have led to the loss of biodiversity (McNeely, 1988). Policy distortions such as these have contributed directly, to increases in deforestation rates in recent decades—not only in developing nations, but in the United States and other developed countries as well (Repetto and Gillis, 1988).

Third, some natural resources, such as biological diversity, are public goods that are used but not owned in the classical sense. Weak ownership, as in the case of nationalized resources or traditionally managed common property in many developing countries, can promote a rush to benefit from what one controls or has access to today, but may not control tomorrow—the "free-rider" problem (NRC, 1986b).

Finally, consumers do not have adequate access to information about the total value of natural resources. Much of this information—particularly regarding nonmarket services such as ecosystem functions, as well as most commodities in the tropics—simply does not exist. When information is inadequate or nonexistent, prices do not reflect resource scarcity.

As resources are depleted for short-term gain, the potential for future use, profit, and development is jeopardized, and when resource depletion involves or leads to reduced biological diversity, even the basic conditions underpinning potential change are compromised. If the present economic system allows only the conservation of what is currently too expensive to exploit (e.g., tropical forests in inaccessible mountain valleys) or too valuable to destroy (e.g., groundwater quality), then the conservation of biodiversity will require either that the present economic system change or that biodiversity be made too expensive to exploit or too valuable to destroy.

The economics of biodiversity conservation raises questions that demand carefully thought-out and often sophisticated answers. Simplistic research will not provide the required information. Research in this area should ideally identify the economic forces leading to the loss of biodiversity within a country; determine the role of international economic institutions and trends in promoting this depletion; elucidate the economic principles operant in cases of successful development and conservation; and develop and test economically viable mechanisms for slowing the rate of resource depletion while stimulating the conservation of biological diversity.

Such research should be multidisciplinary and, in particular, should bring together economists and ecologists. It should also be collaborative, with researchers from developing and industrialized countries participating as equal partners. Because economies operate at various spatial scales, research should occur at project (micro- or mesoeconomic), national (macroeconomic), and global (international economic) levels.

PROJECT- OR COUNTRY-LEVEL ECONOMIC RESEARCH

Ultimately, natural resources, including biodiversity, are managed or mismanaged by national governments or local peoples, depending on the degree of control these groups have over resources and each other. In addition, development agencies concentrate their greatest efforts at the country level. Therefore, project- to country-level socioeconomic research is of greatest importance and urgency. In particular, effective conservation of biological diversity will require more accurate and better-focused information on causal mechanisms, valuation, and incentives and disincentives.

Analysis of Causal Mechanisms

Economic instruments (including rent, taxes, royalties, and concessions; government-financed development; and land tenure systems that depend on landscape alteration) are integral parts of development programs and should be analyzed carefully to determine their short- and long-term effects on the rate of natural resource depletion. Both the local and national economic instruments that lead to the depletion of natural resources and the beneficiaries of such exploitation must be determined. This kind of analysis must occur before or early in the planning of development programs for intervention to be effective.

Examples of the analyses of economic factors that promote the loss of biodiversity in forest ecosystems can be found in Repetto and Gillis (1988). This analytic approach should be applied to other systems, including arid lands, wetlands, freshwater systems, marine and coastal zones, and agroecosystems.

Research Agenda

The following areas of research on causal mechanisms affecting the conservation of biological diversity are recommended for increased attention:

- Determine which (and how) economic instruments used to stimulate the economy are affecting the depletion of biological resources.

Definitions of Rent and Costs

Resource Rent. Rent is a residual or surplus paid to the owners of a resource after payments to all other factors of production are deducted. It is the "price" of the natural resource in situ and is attributable to its scarcity or distinguishing characteristics (i.e., better-quality land or forest can earn more rent).

Production Costs, User Costs, and Environmental Costs. Natural resources, including biodiversity, are undervalued and thus overexploited—not only because *environmental costs* are ignored but because *user costs* are ignored due to a lack of secure, exclusive, and transferable ownership of these resources. Ownership may be held in common, rather than by individuals or agencies (such as government), but the argument is the same.

Environmental costs are those incurred as a consequence of current resource management practices, but not factored in when determining current production costs and benefits. User costs equal the cost of current exploitation in terms of diminished future availability. To the extent that the resource would appreciate over time faster than the discount rate, the resource, or part of it, must be conserved for future use. Under insecure tenure or short concessions, there is no guarantee that the current user of the resource will have access to it in the future; therefore, the user infinitely discounts future benefits (i.e., ignores the user cost of present exploitation of the resource).

- Compare the relationship between the structure (time and cost) of concessions and the willingness of concessionaires to improve or restore conditions. Would tradable concessions improve management?
- Determine the degree to which the lack of property rights affects investment in the development of biological resources by the private sector.
- Examine how price instability in the major export crops affects extraction rates for biological resources.
- Quantify the impacts of different land use practices on long-term economic profitability, both on-site and downstream.
- Determine the economic and environmental trade-offs of annual, perennial, and mixed cropping agricultural production systems in the face of fluctuating markets.

• Determine which groups gain and which are deprived under the current distribution of economic benefits and environmental trade-offs.

Valuation Research

If biodiversity is too valuable to destroy, but the policymaking process does not yet recognize its value, then valuation research must have extremely high priority. Research that establishes the value of biodiversity can improve policy in three ways: (1) Many national and international agencies use benefit-cost analysis to evaluate proposed development projects. More complete analyses that incorporate environmental costs and benefits on equal terms with commercial commodities will serve to eliminate some environmentally devastating projects. (2) Benefits and costs are often considered (although perhaps not as formally as in project evaluation) when policy choices are made. Information on the value of environmental goods and services will tend to encourage the choice of environmentally sensitive policies. (3) Policy becomes effective when it is translated into incentives that direct public or private decisions. Because the choice of appropriate incentives is often empirical, valuation research must underlie that choice.

The valuation of environmental goods and services is usually performed in the framework of benefit-cost analysis. A complete benefit-cost analysis must include all relevant environmental costs and benefits. However, the project analyses traditionally employed by many development agencies pay scant attention to environmental costs or benefits and as a result are biased toward a narrow conception of "development."

A recent study (Peters et al., 1989) shows how merely expanding standard benefit-cost analysis to include the instrumental value of minor products can have significant management implications. In their study of the Peruvian Amazon, the value of harvestable fruits from the forest was calculated on the basis of local market prices and compared to the value of managing the forest for timber or converting it to cattle ranching. The results show that even without including environmental externalities, which would decrease the benefit values for both forestry and cattle ranching, the revenue from extractive fruit harvesting was about double that projected at current market prices for the other two land management options.

Valuation of environmental goods and services is a challenging task, and methods are continually evolving and developing*. Several methods have been recognized in the professional literature and in federal government documents (e.g., see the Principles and Guidelines

---

* Perhaps the most current comprehensive review of valuation methods is Braden and Kolstad (1991). An excellent review of contingent valuation methods is provided in Mitchell and Carson (1989); see also Orians (1991). For application of these methods to water resources, see Kneese and Bower (1968); and Kneese (1984).

Definitions: Valuation Terminology

Valuation. Valuation refers to the process and procedures by which the value of nonmarketed goods and amenities is estimated, so as to be comparable with the value of marketed items.

Benefit-Cost Analysis. Benefit-cost analysis refers to an economic analysis to determine if the estimated present value of benefits from some proposed project or policy exceeds its estimated costs. Benefits and costs are defined broadly to include all the goods, services, and amenities that people care about, whether or not these items are customarily bought and sold in markets.

Use Value. Use value refers to the value of environmental services in use. These services—visitation, recreation, nature study, etc.—have use value just as does the direct harvesting of natural resources.

Existence Value. People may be willing to sacrifice income to ensure the continued existence of natural objects and environments they never actually expect to use. Such objects and environments are said to have positive existence value.

Option Price. Option price is the amount that people would pay in advance to guarantee an option for future use of some environmental amenity whose availability would otherwise be uncertain.

Option Value. Until recently, economists frequently asserted that option price could be broken down into two components: expected value and option value. Expected value was the mathematical expectation of use values in an uncertain situation; risk-neutral people would be willing to pay, in advance, the expected value. Option value was the premium that risk-averse people would be willing to pay in addition to expected value. Recent literature has challenged this formulation, and option value is currently a highly controversial concept.

for Evaluating Proposed Federal Water Resources Projects and the compensation provisions of the Comprehensive Environmental Response, Compensation, and Liability Act).

Briefly, valuation methods fall into three categories. Rather standard market-based economic analyses can be used to evaluate environmental goods that are close substitutes for ordinary commodities. Second, market observations can be analyzed (e.g., via the travel cost method or hedonic price analysis) to infer the value of related environmental goods. Finally, contingent valuation can be used to value environmental goods in simulated choice situations (e.g., contingent markets or contingent referenda). These methods attempt to ascertain the trade-offs that citizens are willing to make between environmental goods and other kinds of commodities. Hundreds of such empirical assessments have been undertaken in the United States and other industrialized countries.

Recently, economists have tried to define a total valuation approach that includes existence value and various kinds of use values in a coherent structure (see Randall, 1991a). A total valuation structure may be developed for situations of uncertain demand and availability, in which case it subsumes the concepts of option price and option value.

Some economists have also begun to explore methods by which the rights of future generations to biological diversity might be better assured. This approach rests on the argument that unless these rights are considered and protected, exercises in valuation will be limited by the current generation's understanding, behavior, and preferences. In effect, the perspective shifts emphasis away from devising means to internalize externalities and toward the transfer of natural assets to future generations. Others have suggested that the same goal can be

Definitions: Valuation Methods

Hedonic Price Analysis. Hedonic price analysis uses economic and statistical analyses to estimate implicit values, including environmental values. Local wage rates and housing prices, for example, may reflect local environmental quality.

Travel Cost Method. The expenditures people make to get to particular sites contain information about the value of the amenities (e.g., recreation, sightseeing, nature study) available at those sites. This method analyzes these travel-related expenditures to estimate environmental values.

Contingent Valuation. Simulated markets or policy choice situations (e.g., referenda) are created, and citizens are asked to report the trade-offs between, for example, environmental goods and income that they would find acceptable. Responses are analyzed to estimate environmental values.

achieved by empowering members of the current generation or existing institutions to serve as guardians for the welfare of future generations or by devising some safe minimum standard (SMS) of resource protection (Norgaard, 1991; Randall, 1991b).

Only recently have these methods and considerations been applied to the valuation of biodiversity in developing nations. Although some obvious impediments exist (they require the use of—or at least familiarity with—well-developed political institutions and market mechanisms, which may be absent in some developing countries), clear signals that such research is demanded in the developing nations will encourage increased effort and lead to significant progress. Valuation of environmental goods in developing nations is a challenging task, but not an impossible one.

Determining the total value for a natural resource is only the first step toward conservation of that resource. This information can be used in project evaluation (where it might serve to eliminate projects that would devastate biodiversity), in policy analysis (where it might provide additional credibility to claims on behalf of biodiversity), and in the development of incentives (where it might capture and distribute values so as to reward actions that enhance biodiversity).

For example, the potential tourism value of a coral reef may be high, but unless there is a tourist industry capable of capturing that value and allocating it not only to hotel developers but also to local people who rely on the reef for their livelihood, the reef may succumb to less environmentally desirable management practices. Similarly, reporting an existence value of $5 million dollars for a Brazilian rain forest may do little to encourage its conservation unless there is a mechanism or institution to capture that value and distribute it to those who will have to forgo the benefits of other management options. The highest values for many biologically diverse and unique ecosystems are likely to be existence and option values—values for what are essentially public goods. Up to this point, however, the conservation and development communities have achieved only limited success in establishing effective institutions to finance and provide public goods in the international context.

Research Agenda

The following areas of research on valuation related to the conservation of biological diversity are recommended for increased attention:

- Review the literature on natural resource valuation methods to identify potentially effective applications in the developing nations.
- Conduct pilot studies, using each of the established methods.

Although contingent valuation is likely to have the broadest applicability, hedonic price analysis and the travel cost method should not be ignored. Efforts to adapt the policy choice referendum form of contingent valuation may be especially rewarding.

- Develop methods, procedures, and guidelines for systematically incorporating biodiversity and other environmental values in routine project evaluations.
- Because the public interest in conserving biodiversity extends beyond national boundaries, research should explore the possibility of worldwide "demand" for conservation (e.g., the degree to which Americans, Europeans, and Asians are willing to pay to preserve Amazonian forest or Arctic ecosystems).
- Determine who captures resource benefits at present. For example, which groups gain direct economic benefits? Do they employ local workers? What percentage of beneficiaries are natives?
- Determine those institutions that can capture the values associated with different land use options and conservation practices. Use gaming methods to determine how modifications of these institutions would change their ability to capture the value of conservation.
- Conduct comparative studies on the value of traditional and nontraditional market uses of natural resources and biologically diverse ecosystems.
- Calculate how the inclusion of environmental externalities in royalties or stumpage fees affects the rate of resource depletion and profits for the owner.
- Identify the international institutions that do or could capture the value of conserving biodiversity.
- Determine whether the value of natural capital increases when it complements man-made capital and vice versa. For example, how does the value of a dam increase when there is a forest upstream? Is the forest worth more when there is a dam at the base of the watershed?
- Undertake research on the intergenerational economics of biodiversity conservation and of sustainable development in general, and on the means by which the rights of future generations might be incorporated into valuation analyses and incentive programs.

Research on Economic Incentives and Disincentives

Incentives and disincentives can sometimes be used to induce governments, local people, and international organizations to conserve biological diversity. To be effective stimuli for conservation, incentives must be applied to all levels of society that affect the rate of resource depletion—community, state, and national. They must also provide benefits perceived as equal to the benefits obtained from resource depletion. To date, perverse incentives, those that induce behaviors that deplete biodiversity, have been much more effective than incentive systems that promote conservation (McNeeley, 1988; Repetto and Gillis, 1988). In fact, some argue that the rate of resource depletion could be curbed drastically by simply removing the incentives that perpetuate inefficiency in production systems.

Positive incentives can be either direct (under which those that forgo the option to deplete are paid directly for conservation) or indirect (under which benefits arise from the installation of conservation policies). A direct incentive may involve, for example, the employment of local people to manage and act as guides in forest reserves. An indirect incentive was put in place, to cite an actual case, when the Yucatan government in Mexico gave local communities the exclusive rights to use and manage a coastal area that was being overharvested for spiny lobster. Once ownership of the resource was in the hands of the community, overharvesting ceased. Similarly, dependence on the use of chemical fertilizers, which can lead to degraded soil structure, accelerated erosion rates, groundwater contamination, and the loss of soil biodiversity, could be discouraged by removing subsidies and adding a small tax to chemical fertilizers while providing a small subsidy for the use of nonchemical fertilizers and for conversion to sustainable agricultural technologies.

Effective disincentives require (1) knowledge of the effects of management activities on the environment, and (2) regulatory and judicial institutions able to design and enforce the disincentive system. Endangered species laws, which fine individuals and impose trade restrictions on nations that permit exploitation of endangered species, provide an excellent example of disincentives. Although some developing nations have environmental ministries and conservation laws, most do not. Those that do often provide inadequate resources to the regulatory institutions or lack the political will to enforce existing regulations.

Successful incentive or disincentive strategies must also be attuned to macroeconomic conditions and able to respond as those conditions change. This is illustrated, for example, in the economics of logging and replanting in Indonesia. The present value of dipterocarp tropical hardwood logs harvested from East Kalimantan forests in Indonesia is $100 per cubic meter, yielding on average 40 cubic meters per hectare with a total value of about $4,000. The cost of reforesting cutover areas averages $1000 per hectare. Until recently, however, the refundable replanting fee that loggers had to pay was only about $160 per hectare ($4 per cubic meter × 40 cubic meters). As a result, loggers chose to pay the fee rather than replant and receive the refund. The fees were not sufficient to encourage replanting of the area harvested. Recently, the government increased the replanting fee to $7 per cubic meter and made it nonrefundable. Fees are earmarked for rehabilitating degraded areas (such as *Imperata* grasslands) while the concessionaires are still responsible for replanting their concessions.

A relatively new disincentive system involves *environmental bonding*, wherein a company or institution wishing to exploit a natural resource or utilize an environmental service posts a bond. The size of the bond must be sufficient to cover the costs of lost production and site rehabilitation should the worst possible environmental disaster associated with that exploitation occur. For example, a logging company might be required to post a bond large enough to cover the costs of reclamation and revegetation of the area it plans to cut. The company would be eligible for a refund of part of the bond for each year in which the worst-case scenario did not transpire.

Environmental bonding places the burden of proof that no harm will occur on the resource exploiter. It ensures that sufficient resources will be available for rehabilitation should disaster strike, and it encourages the exploiter to conduct research to decrease the degree of uncertainty associated with the occurrence of the worst-case scenario as a justification for a reduction in bond size (Costanza and Perrings, 1991). To date, bonding has been tested sparingly in industrialized countries and remains an untested disincentive system for conservation in developing nations.

Research Agenda

The following areas of research on the effect and application of incentives and disincentives to issues of biological diversity are recommended for increased attention:

- Evaluate the effectiveness of existing disincentive systems and laws.
- Document and explain cases in which incentive systems have successfully conserved biodiversity.
- Determine how to adjust incentive systems to achieve a more efficient and sustainable allocation of resources.
- Determine how incentives can be used in biodiversity restoration efforts in degraded systems.
- Identify institutional constraints on the implementation of incentive systems at the local and national level, and develop strategies for the elimination or mitigation of these constraints.
- Determine whether price fluctuations in agricultural commodities function as perverse incentives for agricultural expansion into natural areas.

INTERNATIONAL-LEVEL ECONOMIC RESEARCH

The areas of economic research noted above—causal mechanisms, valuation, and incentives/disincentives—pertain primarily to national and local level activities. However, many of the forces that profoundly affect local practices and national economic policies, and hence the depletion or conservation of biodiversity within a country, are transnational in origin. The relationship between these forces and natural resource exploitation within a country is poorly understood. Therefore, more research is needed that focuses on the interface between national and international economic factors, institutions, trends, and impacts. The following areas of international-level economic research on issues relevant to the conservation of biological diversity are recommended for increased attention:

- Determine the impact of inflation, tight money, external debt, and market instability on the rate of natural resource depletion.
- Evaluate how these factors affect the efficacy of incentive and disincentive systems.
- Examine if and how trade barriers and most-favored nation status impact exploitation of natural resources.
- Determine the effects on internal economic stability and national support for conservation through debt-for-nature trades and other financing instruments that link national and international financial organizations.

GLOBAL ECONOMIC RESEARCH

Some issues and actions that result from the interrelationship between economics and biodiversity can be addressed only at the global level. For example, many developing nations may simply be unable to fund conservation activities at the level required or to support the types of direct incentive programs that would be effective in conserving biological diversity within their boundaries. Biodiversity is a global resource, and multinational and multi-institutional efforts will be required to supplement and complement attempts at the national or project level to conserve it. Therefore, more research is needed that focuses on macroeconomic forces operating on a global scale. The following research topics that address the conservation of biodiversity on a global scale are recommended for increased attention:

- Identify the worldwide constituency for conserving biodiversity.
- Determine how different types of institutions compare in their

efficiency and adequacy in handling the uncertainty associated with environmental manipulations.

- Study the impacts of rapid growth of international financial flows and markets on the environment in general, and on biodiversity in particular, in developing nations. For example, what effect do reforms mandated by the International Monetary Fund—including structural and sectoral adjustment programs and other liberalization reforms—have on biological resource depletion in debtor nations?
- Document how policies and management for macroeconomic objectives can lead to mismanagement of economies based mainly on biological resources, particularly local microeconomies.
- Compare the kind and degree of impacts associated with free-market and planned economies with respect to natural resource depletion.
- Identify the global economic factors critical to sustainability in developing countries, and their relationship to social and environmental factors. Especially important in this regard are case studies in sustainable and unsustainable development. This may lead to the formulation of alternative resource indicators that can be compared to prices in monitoring changing resource scarcity.

OPPORTUNITIES FOR ACTION

The long-term conservation of biological diversity in developing countries will require biological and social research of the sort described elsewhere in this report. Most of the economic research described in this chapter, however, is intended to stimulate immediate action to stem the loss of biological diversity. Innovative valuation research can help to ensure that environmental goods, services, and amenities do not go undervalued, and hence underrepresented, in the economic assessment of development projects and policies. Causal mechanism analysis and research on economic incentives and disincentives can lead to the rectification of current policies and the development of constructive new measures. This action-oriented approach should guide all those who undertake the conservation of biological diversity from the economic perspective, even as the natural sciences and the other social sciences continue to provide the basic facts concerning life's diversity.

4

Biodiversity Research: the Cultural Context

Human beings have occupied this planet very thoroughly for thousands of years, and few "natural" habitats remain. The landmasses of the Earth are largely covered by mosaics of habitats that have been altered to a greater or lesser degree by a rich diversity of human action. Some of these areas are occupied by people who have developed approaches to managing the resources of their local ecosystems in a sustainable fashion, while others are occupied by people whose activities have recently altered those systems to such a degree as to call their sustainability into question.

For most of these thousands of years, anthropogenic environmental impacts were largely confined to local areas, and took place at a rate that allowed societies and their resource base to adjust to one another and maintain themselves in a general state of equilibrium. Relatively sparse human populations; subsistence technologies; local control over resources; land ownership by clans, ancestors, or lineages rather than individuals; and other social mechanisms protected much of the natural world from the massive degradation that is now leading to widespread loss of biological diversity.

This, however, was never an ideal world; the "Garden of Eden" vision contains its weeds. The equilibrium was not a static state. Where human hunters moved into new habitats filled with naive game animals, major extinctions have occurred; the Americas, Australia, Madagascar, and New Zealand provide well-known examples (Martin, 1984). Humans are also implicated in the extinction of some 90 percent of the endemic mammalian genera of the Mediterranean after the development of agriculture (Sondaar, 1977). Ancient human settlements occasionally visited deforestation, desertification, and salinization upon their lands, and eviction upon themselves, in the course of their development.

Although hunting and agricultural societies have long had the power to transform their environments and drive species to extinction, it remains a valid observation that the natural world at the dawn of the industrial age was characterized by highly diverse ecosystems and supported highly diverse human cultures. In the past few generations, however, as fundamental ecological changes have taken place, the world's heritage of natural and cultural diversity has been diminished.

DIVERSITY AND DEVELOPMENT

These shifts have both caused, and been caused by, social changes. The world's collection of highly diverse cultural adaptations to local environmental conditions has begun to be replaced in many locations by a world culture dominated by very high levels of material consumption. Economic growth based on the conversion of fossil fuels to energy has greatly expanded international trade. Improved public health measures have spurred a rapid expansion of human numbers, requiring new approaches to resource management. These approaches have overwhelmed the conservation measures (formal or informal) of local communities, bringing overexploitation and poverty to many rural communities, and great wealth to cities and certain individuals, as urban elites in both industrial and developing countries control policies in such a way that primary productivity is very poorly rewarded.

Technological innovations have tended to promote exploitation of biological resources and to weaken traditional management systems, especially when a dominant group moves into a region occupied by technologically less advanced groups. The dominant society has the option of moving on to fresh resources when an area is exhausted, and it derives no particular advantage from adopting traditions of sustainable use. Its members are able to earn virtually all the immediate cash benefits of a forest, for example, but pay almost none of the long-term environmental costs. In addition, they capture a very small fraction of the potential cash benefits from the forest through short-term exploitation, rather than greater income from judicious long-term management in cooperation with local peoples.

At the same time, the subordinated groups lose any advantage from traditions of conservative use that might have been favored in times when they could exclude other groups from their territory. These traditions evolved when costs and benefits were internalized in the decisions made by communities, but as local peoples have had to assume the higher environmental costs of resource degradation, often their only rational response has been to join the exploiters in seeking greater benefits as well. This is the real tragedy of the commons: traditional management systems that were effective for thousands of years become obsolete in a few decades, replaced by systems of relentless exploitation of rural people and rural countries, those who depend on primary productivity. Exploitation brings short-term profits for a few and long-term costs for many throughout the world.

Diversity, both biological and cultural, is a casualty of this process and of the operant development paradigm behind it. In the push, for example, to modify local farming systems to accommodate modern Western technologies and to put in place the institutions required to manage these technologies, little heed has been given to the complexity of systems such as tropical rain forests or their indigenous peoples. As more consumptive exploitation of biological resources occurs, cultural diversity is often reduced, for two main reasons. First, a significant component of cultural diversity that enables people to earn a living from the local biological environment is no longer functional; second, subordinated groups must often imitate the culture of the dominant group, thereby losing a substantial portion of their cultural identity and, hence, their diversity.

The World Commission on Environment and Development (1987) described the process:

> Growing interaction with the larger world is increasing the vulnerability of isolated groups, since they are often left out of the processes of economic development. Social discrimination, cultural barriers, and the exclusion of indigenous people from national political processes make them vulnerable and subject to exploitation. Many groups become dispossessed and marginalized, and their traditional practices disappear. They become the victims of what could with justice be described as cultural extinction. . . . It is a terrible irony, that as formal development reaches more deeply into rain forests, deserts, and other isolated environments, it tends to destroy the only cultures that have proved able to thrive in these environments.

As noted in the previous chapter, the desirability and sustainability of the standard development paradigm are being challenged by growing numbers of people in the developed and developing nations, and within the international development community. This comes as a variety of profound impacts of environmental exploitation—including not only the loss of biodiversity that is the focus of this report, but also associated concerns about human population growth, depletion of the Earth's ozone shield, and possible changes in climate resulting from anthropogenic emissions of greenhouse gases—suggest the inevitability of changes in the way humans relate to the global environment. In this context, the need for more sustainable development strategies and technologies has become apparent, and awareness is growing that

The Demographic Context

Since 1950, the world's human population has more than doubled, from 2.5 billion to its present level of 5.4 billion (PRB, 1989). Of these people, 77 percent live in developing countries and control about 15 percent of global wealth. One billion people will be added to the developing countries in the 1990s, at the rate of more than 90 million people (the population of Mexico) per year. At the same time, topsoil is being lost from the world's agricultural lands at the rate of about 25 billion tons per year, and about 20 percent of the total has been lost since 1950. This trend is essentially irreversible: in the tropics, it takes about 100 years to build an inch of topsoil, and little additional prime land is available for cultivation. We are already managing global ecosystems; our challenge now is to manage them well and not remain passive while allowing them to deteriorate.

traditional community-based resource management systems have much to offer as a new development paradigm emerges.

LOCAL KNOWLEDGE AND BIODIVERSITY

Only by determining why people do what they do can we understand the relationships between people and their biological resources. Which species are most valued for food, medicines, construction, or ceremonial uses, and why are they treated in certain ways? Why do people locate their houses in certain ways in relation to the landscape? Why do subsistence farmers conserve low-yielding local cultivars? Underlying all of these questions are behavioral factors that must be understood, because they determine how people will respond to external pressures that influence their use of resources.

The social sciences can mediate between indigenous resource use patterns and other institutions by examining indigenous knowledge to understand the problems of biodiversity loss and the methods for biodiversity conservation. The social sciences help identify not only the components of the local knowledge system, but also the timing, behavioral processes, and structural characteristics that result in the conservation or reduction of biological diversity.

Recently, some individuals in the development community have argued that local knowledge can point toward new types of sustainable agriculture, natural resource management, and conservation (Peters et al., 1989). They propose that greater benefits can be gained by managing a rain forest for sustained extraction than by converting it through more intensive methods. Furthermore, they suggest that the local knowledge system may yield new ideas about the conservation and management of valuable products and identify valuable new biotic resources for utilization. This point of view constitutes a welcome departure from traditional thinking about local knowledge, but may actually encourage the depletion of biological diversity if it seeks merely to identify those efficiently marketed species and goods known to local peoples and fails to comprehend the complete contextual system in which local cultures operate.

As development agencies have come to understand the need to conserve biodiversity and the complexity of the ecosystems and cultures that this task entails, they have recognized the vital role that local peoples—defined variously as indigenous, tribal, traditional, or subsistence—must play in the conservation process if it is to succeed. Yet this recognition has rarely been extended to include the belief that local people can contribute more to the conservation process than simply their acquiescence. Why?

Local people, those who have developed apart from the dominating society that is penetrating their area, play an important role in maintaining biodiversity. Little is known about the resources many of them use and why they use them. Less is known about why they shape ecosystems as they do or about their concept of landscape and the institutions used to enforce such concepts. Local people are expected to be what the dominant society anticipates; it is assumed that local knowledge does not exist or will be beneficially replaced from without. In reality, however, local knowledge not only exists but may make important contributions to the conservation of biodiversity.

Nature of Local Knowledge

Local people, depending on a number of historical, social, and ecological factors, amass an extraordinary store of knowledge about the local natural resource base. Some groups not only construct taxonomies of plants and animals based on useful characteristics, but also compile information on species abundance and distribution (Berlin et al., 1974). Many of these taxonomies are extensive. The Maya of southern Mexico and Guatemala, for example, include not only plants and animals, but soils and ecological communities and successions in their classification schemes (Gómez-Pompa, 1987). The Huastec Maya include 950 species of plants in their classifications; this represents 89

Traditional Management Patterns

Many local peoples have developed complex management patterns for biological resources. The Kayapo of Brazil manage more than 600 plant species in a complex agroforestry system that is intensive, diverse, and sustainable. Peruvian pastoralists rotate animal grazing in high-elevation pastures to maintain mixed-species pastures and thereby ensure food for their flocks even during periods of climatic stress. Australian aborigines had an elaborate fire management regime that controlled the frequency of severe fire and simultaneously maintained high levels of productivity and biodiversity in the outback. Farmers in centers of crop diversity have nurtured a considerable degree of variety in their crop germ plasm. Andean farmers, for instance, keep thousands of potato varieties, adapted to a wide range of edaphic, topographic, climatic, and altitudinal conditions (Brush et al., 1981).

percent of the local flora, 70 percent of which were used in one way or another (Alcorn and Hernandez, 1983). Since local taxonomies are based on use, they are a source of basic botanical information and they also suggest species that might be developed into market commodities.

These examples show that local people identify and classify useful plant and animal species, describe ecological communities in an environmental context, and test and evaluate species for their useful potential. Using this knowledge, they design, test, and develop resource use patterns, establish microenterprises and markets, and develop mechanisms for transferring knowledge from one generation to the next.

Local resource use patterns, then, are not static, outmoded, or unsophisticated, but neither are they suitable to institutionalization in the manner generally promoted by development agencies. There is no single traditional knowledge system that can be used to preserve biological diversity in all rain forests or on all semiarid rangelands, but no modern technological system can perform this function either. The management systems that local people use are based on, and take advantage of, the biological diversity of particular locales. Although they are believed to be suitable only at relatively low population densities, they are critical sources of insight into agroecological principles that can be incorporated into more complex intensive systems. In short, we need to understand the diverse ways in which rural people relate to their environments and the degree to which such relationships are sustainable.

Tropical Technologies

Shifting Agriculture

In many parts of the world, large numbers of people farm small areas of land that fulfill their needs, without regularly using modern agricultural inputs such as chemical fertilizer, high-yielding crop seeds, or commercial pesticides, and they harvest and store their grain using techniques that have been employed for generations. In Africa, for example, the World Bank (1989) estimates that the average use of chemical fertilizer is less than 10 kilograms per hectare, compared with about 90 kilograms per hectare in China and India.

Most of these farmers employ some form of shifting (or swidden) agriculture. In this traditional system, the farmer clears land that has been left fallow for anywhere from 6 to 25 years (Christonty, 1986), uses or sells the valuable woody material for construction or charcoal, and burns the rest in situ to provide nutrients for his next crop. The system depends on natural or managed revegetation to restore fertility to the ground when it can no longer support crop production. How long and how much the land will produce are functions of the inherent fertility of the soil, drainage and leaching rates, human management of vegetation, the degree to which animals are used in plowing or otherwise contribute their manure, and the rate at which the area is encroached by weeds and other pests.

The interplay of these factors determines when the farmer will move to another plot. In some parts of Ethiopia with volcanic soils, farmers have been able to produce between one-half and one ton of grain per hectare every season on the same land for as far back as memories stretch, perhaps hundreds of years, so that shifting is not imperative. In Ghana, by contrast, on well-drained savannah soils, only two or three crops can be produced before the soil is exhausted; weeds make further use impractical until it has fallowed, preferably for 15 years (now often reduced to 5 years because of lack of alternative available land, and therefore with much less recovery and buildup of fertility).

In its traditional form, shifting agriculture depends on the biological diversity of surrounding areas to sustain the natural cycle of clearing and revegetation. In recent decades, however, fallow times have been shortened as a result of population increase (due to both high birth rates and immigration by nonnative farmers), the resultant pressure on land, ineffective pricing policies, and faltering industrial development.

This has several consequences. Restoration of the soil through the action of deep-rooted trees and shrubs, particularly leguminous species that fix atmospheric nitrogen, may be interrupted. The amount, diversity, and quality of biomass produced are lower; consequently, the amounts of biomass or nutrients available for the next cycle are reduced. The process of succession, in which different plants take advantage of different opportunities presented during the successive stages of regrowth, may not continue sufficiently for some species to mature; hence species and biodiversity may be progressively lost. The reduction in diversity often means special losses to the local people, who may depend on regrowing areas for supplements to their diets (seasonal fruits, nuts, small animals) and for a wealth of useful natural products (herbs, dyes, tannins, creepers for cordage, leaves for thatch, and traditional remedies for a wide variety of ailments).

The modernization of agriculture, including the intensification of fallow management and use, has generally been pursued through development schemes that attempt to substitute technological resources for natural resources. This may succeed in helping the farmer farm the same land continuously by using fertilizer, pesticides, and other purchased inputs, but it is seldom able to compensate for or to restore the surrounding diversity of biological material. It brings with it the susceptibility of monocultures to any number of problems, including decreased resistance to pests and drought, and increased dependence on the affordability and availability of hybrid seed, fertilizer, pesticides, and other inputs. Meanwhile, knowledge about the broad range of natural products from the "bush" is dwindling as the amount of bush itself dwindles and as children no longer absorb the lessons of their elders.

Intensive Agricultural Systems

Two types of intensive systems dominate production agriculture in the lowland humid tropics: swamp or paddy rice production systems and perennial tree crop plantations. Swamp rice cultivation is a product of the evolved complex system of land and water management that sustains huge populations in East, South, and Southeast Asia, and has been copied successfully in parts of Central and South America. It involves backbreaking work, and the social organization of communal labor that land preparation, transplanting, irrigation, and harvesting entail has limited the successful development of analogues elsewhere. In Africa, local waterborne diseases—especially filariasis, river blindness, and schistosomiasis—make year-round cultivation with irrigation singularly difficult, although upland rice cultivation has been more successful.

Lun Dayeh of Northeastern Borneo

The Lun Dayeh of East Kalimantan in Indonesia practice both shifting and permanent field irrigated cultivation of rice and a number of subsidiary crops, and are noted throughout much of Borneo for their abundant annual rice harvest. The success of Lun Dayeh farming has been attributed by many not familiar with the group's homelands to extremely favorable natural conditions for agriculture, particularly fertile soils. A closer examination, however, shows that the Kerayan Subdistrict, where most of East Kalimantan's Lun Dayeh live and farm, is a highly varied region, blessed only in very limited areas with exceptionally good conditions for rice production. The success of Lun Dayeh agriculturists is more reasonably attributed to their detailed knowledge of the environments they exploit, sound judgments of where to site their fields, and good management practices.

Like shifting cultivators and other traditional agriculturists throughout the world, the Lun Dayeh of East Kalimantan's Kerayan Subdistrict weigh a number of environmental variables before selecting a site for a new agricultural field. Although the environmental variables most important to swiddeners—including the composition and height of the vegetation covering the area—are taken into account, Lun Dayeh permanent field cultivators tend to be more concerned with other environmental characteristics including slope, water quality and availability, drainage, and soil quality both on and below the surface. Moreover, they attempt to judge the long-term acceptability of the site. Therefore, continuing observation of an area, often for years before it is first used, is not unusual. Although the criteria for an acceptable site for an irrigated farm field are many, Lun Dayeh cultivators assume that some changes will be necessary to bring a field to the condition required. Therefore, a most important consideration is the amount of labor that will be needed to render a site acceptable for pond-field farming. Because population density in the Kerayan is low and almost ideal areas are still available, Kerayan Lun Dayeh farmers continue to reject any site that requires major improvements such as significant terracing work, filling in sites that tend to flood too deeply, or digging long irrigation canals. As most easily worked areas are claimed in the future, what is judged an acceptable site will certainly change, and the construction of ever more labor-demanding engineering works will be considered necessary and reasonable.

Another facet of this process is found among the Ifugao of Luzon. According to the reports of Conklin (1980), in choosing a site, the Ifugao, consider the incline of an area the most important criterion. Because flat or gently sloping lands are no longer available in Ifugao, an acceptable slope is far more steep than any site that the Lun Dayeh would attempt to exploit. Also important are the availability of a water supply upslope from the chosen site and the location of materials necessary for building terraces in the vicinity of the selected area; these include wailing rock, rough fill, and topsoil. In the Ifugao case, cultivators expect not to find, but rather to create, most of the conditions necessary for pond-field farming; their fields, in contrast to those found in the Kerayan, can be considered largely artifacts, because the soil, including subsoil and the entire surface of each terrace, has been created with human labor.

The willingness and ability to alter environmental conditions, which in large measure distinguish Lun Dayeh pond-field farmers from their swiddening counterparts, are considerably more developed among wet rice farmers such as the Ifugao whose homelands offer fewer unexploited ideal sites for irrigated cultivation. A comparison of labor needed to create pond fields in different areas would be interesting but difficult because of the great variability of sites available in each location. However, initial pond-field creation costs, as well as subsequent labor for maintenance in regions such as Ifugao, surely far surpass those found in the Kerayan. With ever-increasing alteration of the original environment, the need for human labor in maintenance increases as well. The Lun Dayeh, therefore, in seeking out areas that naturally approach most closely the conditions desirable in an irrigated pond field are minimizing not only initial construction costs but also subsequent maintenance (Padoch, 1986).

Perennial plantations of tree crops have been extremely successful throughout the tropics, to the extent that overproduction and control by industrialized countries of the markets for rubber, coffee, cocoa, tea, copra, palm oil, and other commodities have led to steadily falling prices. Of special interest now are possibilities for diversifying forest productivity, based on an understanding of traditional systems practiced in Southeast Asia and South and Central America. In diversified plantations, a mixture of useful species is maintained, providing the farmer with a broad range of products, including tropical hardwoods; fruits and other edible products; oils and resins; livestock and wild game; and rubber, cloves, cinnamon, or other cash crops. Recent studies of the Peruvian Amazon have shown that the value of products obtained from extractive reserves of forest can exceed the value of production of food crops or ranching on converted forest areas (Peters et al., 1989).

Extensive Systems

Pastoral nomadism is another type of cultural adaptation to drought or aridity and, in many cases, to the health and pest problems associated with settled agriculture. Arid lands support large populations of people and animals—an estimated 135 million, or about 20 percent of the world's population, and most of the 3 billion head of domesticated livestock (NRC, 1990). Moisture limits the productivity of the systems, which are necessarily much more extensive than the systems of tropical lowlands and forests.

There are two general systems of rangeland utilization: systems that use the land to produce goods that are removed or exported (ranches), and those that chiefly provide subsistence for people associated with livestock and wildlife populations (indigenous pastoral systems). Contrary to popular belief in industrial nations, pastoral systems are not necessarily less productive than ranching systems. African pastoral systems, for example, are often as productive as market-oriented ranching systems in comparable areas in terms of protein produced per unit of land utilized.

Most ranches are privately owned and characteristically use both the investment of capital and various management techniques on large areas of land to increase livestock production. Unlike pastoral systems, labor inputs are low. Hence, ranching often produces more protein per hour of labor than does pastoralism. On the other hand, ranching requires vastly greater inputs of energy, and the expenses incurred in connection with fencing, water development, brush control, revegetation, grazing management, and selective breeding are substantial.

Rangeland ecosystems, particularly those in arid and semiarid regions, are highly susceptible to degradation. In many regions, degradation is chiefly a result of changing herd composition and overstocking. Particularly noteworthy since the advent of the colonial period has been a shift in herd inventories favoring cattle, a form of livestock poorly adapted to dryland ecosystems, at the expense of well-adapted and less environmentally destructive forms, such as

Pastoral Systems

Pastoral systems represent the principal form of rangeland utilization in Africa and Asia. They involve significant social adaptations to the movement of livestock or wildlife from one location to another in relation to the availability of forage and water, and the avoidance of diseases such as trypanosomiasis (sleeping sickness) which in Africa is endemic in wetter areas where the trees that are the habitat of the tsetse fly vector are found. The rangelands utilized are seldom privately owned, and mechanical or chemical inputs are seldom prominent. The systems are labor intensive. It has been estimated that livestock and wildlife support some 30 to 40 million pastoralists, and the animals and animal products associated with pastoral systems are critical to millions of other individuals in settled communities (World Resources Institute, 1987).

The importance of livestock in pastoral systems exceeds their value as sources of milk, meat, blood, and hides. Livestock often represent a means of accumulating capital and, in some societies, are associated with social status. They are assets that can reproduce and can be liquidated should cash be required. In addition to supporting livestock, rangelands serve as sources of other significant economic products: bushmeat, fruits, berries, nuts, leaves, flowers, tubers, and other food for human populations, as well as medicinal plants, building materials, thatch, fencing, gums, tannin, incense, and other products important to the economies of rural populations (Sale, 1981; Malhora et al., 1983, National Research Council, 1983).

The importance of rangelands as sources of bushmeat and vegetable foods for human populations deserves special attention. These foods are derived from species that are well adapted to the environmental peculiarities of the regions in which they are found. Hence, such foods are often available in the event of crop failure or substantial loss of livestock. Even during periods with average rainfall, satisfactory crop yields, and herd stability, such foods constituted a significant part of local diets. Indeed, in many societies, the offtake of wildlife from rangelands exceeds that of livestock in importance. In 1959, for example, the sedentary and pastoral peoples of the Senegal River Valley in West Africa relied on fish and wildlife for more than 85 percent of the meat they consumed (Cremoux, 1963); native plants were of equal or greater importance. Since that time, widespread environmental degradation has dramatically reduced the availability of natural products associated with local coping strategies and has correspondingly increased the vulnerability of rural populations (NRC, 1983). In most instances, the degradation is a result of breakdowns in the traditional resource management systems that for centuries have maintained an equilibrium between environmental systems and human activity (NRC, 1986b).

---

camels, because the former are more marketable in the context of the new economic order (Chassey, 1978). In the West African Sahel, for example, colonial policy resulted in an almost fivefold increase in the cattle population between 1940 and 1968.

Agricultural expansion has also contributed to the degradation of tropical and subtropical rangelands. In drylands, agricultural expansion results in increased pressure on rangelands because conversion of the more productive forage reserves to cropland forces pastoralists to overgraze the remaining land base (Thomas, 1980). Moreover, grain crops deplete soil nutrients at a rate 30 times greater than the rate of nutrient loss in a properly stocked range ecosystem. The cost of replacing the lost phosphorus, potassium, nitrogen, and other nutrients is generally prohibitive.

In many regions, high levels of sustained use pressure have eliminated the more palatable plant species. In dryland ecosystems, plant growth is relatively slow. When aerial biomass is consumed by foraging livestock, many plants respond by transferring nutrients from their roots to produce new leaves, which results in reduced rooting. Reduced rooting, in turn, decreases the ability of the plant to absorb moisture and nutrients even during rains. As the more palatable species are weakened with continuing high levels of use pressure less palatable species of undesirable shrubs, grasses, and forbs become dominant. As these species are overgrazed, the land surface is exposed to further, more severe, degradation. In the drylands of Africa and Asia, cattle have been particularly destructive. Unlike camels, goats, and most native herbivores, which are predominantly selective browsers, cattle are grazers; they therefore increase the pressure on perennial grasses and often eliminate them, causing ecological deflections toward ephemeral annual grasses and relatively unproductive trees or shrubs, such as *Calotropis procera* (Gaston and Dulieu, 1976).

The reduction or elimination of vegetative cover, in combination with trampling and compaction of the surface by livestock, reduces

Crop Diversity

Among *the Lua (Lawa) of northern Thailand*, about 120 crops are grown, including 75 food crops, 21 medicinal crops, 20 plants for ceremonial or decorative purposes, and 7 for weaving or dyes. The fallow swiddens continue to be productive for grazing or collecting, with well over 300 species utilized. The most important crop is upland rice, and it is not unusual for 20 varieties of seed rice to be kept in a village, each with different characteristics and planted according to the soil, fertility, and humidity of the fields.

The *Hanunoo of the Philippines* may plant 150 species of crops at one time or another in the same swidden. At the sides and against the swidden fences grow low climbing or sprawling legumes—asparagus beans, sieva beans, hyacinth beans, string beans, cowpeas. Toward the center of the swidden, ripening grain crops dominate, but many maturing root crops, shrub legumes, and tree crops are also found. Pole-climbing yam vines, heart-shaped taro leaves, ground-hugging sweet potato vines, and shrublike manioc stems are the only visible signs of the large store of starch staples building up underground, while the grain crops flourish a meter or so above the swidden floor before giving way to the more widely spaced and less rapidly maturing tree crops. A new swidden produces a steady stream of harvestable food in the form of seed grains, pulses, sturdy tubers, bananas, spices, and many others.

Among the *Tsembaga Mareng of Papua New Guinea*, each field contains some 10 to 15 major crops, plus dozens of minor crops, spread seemingly at random through the field. This intermingling discourages plant-specific insect pests and takes advantage of slight variations in garden habitats. It also protects the thin tropical soil and achieves a high degree of photosynthetic efficiency.

The *ribereños of the Peruvian Amazon*, nontribal descendants of the indigenous inhabitants, have adapted traditional swidden-fallow agroforestry patterns to the market economy. They largely retain the age-old cyclic system, following a diverse planting of annual and semi-perennial crops with a mixture of perennial tree crops and forest species. However, they have also altered the traditional system in several crucial ways. At each stage of the cycle, which may be 25 years or longer, the particular choice of crops and methods is determined both by the opportunities the markets present and by the subsistence needs of villagers. Households in some ribereño villages gain yearly incomes of $5,000 or more from their agroforestry fields while still maintaining a high diversity of crops and the productive potential of their lands.

---

infiltration and permits the mobilization of soil particles subject to transport by overland flow. This results in depressed groundwater tables and increased soil erosion. Surface exposure and the reduced organic content of soils also result in altered soil-water relationships and greater variation of soil temperature. The altered soil ecology adversely affects important soil microorganisms such as the rhizobial bacteria responsible for nitrogen fixation in acacias and other leguminous genera. In turn, nutrient regimes are affected and further loss of soil structure results. Altered soil ecology directly eliminates additional plant species and frustrates regenerative processes in others. More losses occur through disruptions in various biological dependency and affinity relationships. Environmental degradation both reduces range-carrying capacity for livestock and affects wildlife populations through habitat modification.

The effects of rangeland degradation often extend well beyond the rangelands themselves. Dust originating in degraded rangelands is transported by dry-season winds to distant areas, causing annoyance, health hazards, and costly interruptions in air and ground traffic. The rapid release of runoff in degraded rangelands following rains contributes greatly to destructive flooding in downstream lowlands, and sediment entering drainage systems in degraded rangelands shortens the useful life of reservoirs and irrigation systems.

Less obvious effects would include the impact of rangeland devegetation on climatic regimes. For example, it is now widely believed that precipitation is strongly influenced by biogeophysical feedback mechanisms (Charney, 1975). Further, it is now believed that precipitation levels are strongly influenced by soil moisture locally released into the atmosphere through evapotranspiration. Hence, reduced vegetative cover and decreased soil moisture would result in reduced local precipitation. Finally, losses of vegetation affect surface roughness in the atmospheric boundary layer. Surface roughness contributes to the destabilization of moisture-laden air masses, thus encouraging precipitation. Devegetation also reduces carbon dioxide uptake in the planetary biomass. The greater concentration of carbon dioxide in the atmosphere contributes to global warming, causing changes in atmospheric circulation and rising sea levels through the melting of continental ice sheets.

Historically, attempts to transfer experience gained in the management of North American or European rangelands to the management of tropical and subtropical rangelands have been unsuccessful (Heady and Heady, 1982). In managing tropical and subtropical rangelands, it is important to characterize carefully the physical system being managed in order to better understand the biological potential of the system and ensure that critical ecological processes are restored and maintained. It is equally important to relate efforts in range improvement to the needs, knowledge, adaptations, and capabilities of local populations, as well as to the broader economic and political contexts of such efforts. The widespread belief that pastoral systems are simply artifacts of the past requires reexamination. The view that range improvement in the tropics and subtropics should focus narrowly on the increased per unit productivity of selected forms of livestock, usually cattle, at the expense of the biological diversity basic to the maintenance of local coping strategies and economies should similarly be reexamined.

Tropical Forests

Traditional tropical forest technologies—cyclic agroforestry, intercropping, and home gardens—promote the conservation of biodiversity and offer insights into useful ecological associations. Local people are often excellent resource managers when they are allowed to manage their own resources for their own benefits. Development tends to encourage them to change their traditional ways of life and often to become more exploitation minded, including converting very complex multispecies agroecosystems into monocultures, which often encourages overexploitation of the system. The key research problem is to identify how the more complex traditional systems can be adapted to modern needs, for example, to support a growing population and use less labor-intensive methods of cultivation while still retaining the biological diversity of both the agroecosystems and the surrounding lands.

RESEARCH ON LOCAL KNOWLEDGE

As noted above, the cultures that developed and maintained local knowledge, and the systems that sustained productivity and diversity over many generations, are rapidly changing. Local knowledge is being displaced by technologies that have not demonstrated their sustainability or their long-term contributions to society. As development agencies seek to understand the traditional forms of management, research must seek to identify the nature of this productivity and sustainability from the perspective of the cultures in which they evolved, and it must do so before this knowledge is lost.

Fundamentally, *research should provide information on local resource use practices.* We need to compile information on local populations and particular resource management patterns (including the uses of flora and fauna) that exemplify sustainable relationships between people and the environment. An inventory of this knowledge should be compiled, highlighting specific features that can contribute to conservation and development, with special attention given to the identification of endangered local resource use patterns. Educational agencies should be assisted in introducing elements of local conservation knowledge and practice into appropriate curricula, and educational activities should be undertaken to encourage interest in traditional knowledge and its practitioners. To the extent possible, the value of local management systems should be demonstrated by indigenous people themselves.

Research should promote the application of local knowledge to modern resource management, and vice versa. An exchange of knowledge and methodologies would foster greater mutual understanding between indigenous peoples and conservation scientists or managers. To facilitate such consultation, indigenous peoples may need training in the approach and techniques of conservation science. Scientific investigators and researchers should include indigenous coinvestigators in all phases of their research design and implementation, with the objective of establishing networks for the long-term exchange of information and learning. Financial and technical resources should be made available to enable indigenous people to conduct their own research. The aim should be to create an indigenous scientific community that includes both locally evolved and externally acquired expertise, skills, and procedures.

Based on this information, development agencies would be able to *design projects that benefit indigenous people and that benefit from local knowledge.* The agencies should identify opportunities to demonstrate how local knowledge can be combined with science in designing systems for sustainable resource use and developing such projects for external funding. Based on a review of development projects that have involved indigenous people, a workshop should bring together experts to develop guidelines that agencies can use to enhance the design, implementation, and monitoring of their development projects.

To promote the idea that local knowledge and practices (such as customary law) remain relevant for contemporary natural resource management, especially in terms of the scientific insights they provide, the *rationale for examining local knowledge and practices should be communicated to professional groups*. Traditional land tenure arrangements should serve as a basis for planning and executing conservation projects, as well as projects more directly concerned with food and materials production. Marine conservation and inshore fisheries development programs should be based on established rights and tenure systems, and should incorporate local ecological and management knowledge. Where traditional tenure systems appear to be inadequate for markedly changed conditions (in the case of greatly increased human population or resource degradation), new systems should adapt the best features of the old.

Priority Groups for Research

There are, at minimum, hundreds of different ecosystem types, thousands of ethnic groups, and between 10 and 25 million species in the developing nations. Clearly, studying all existing or possible resource use patterns, traditions, combinations, and relationships is impossible. A selection of people and places is required. Of highest priority are those use patterns and knowledge systems that are changing most rapidly or disappearing. The following are recommended for closer study:

- Foragers and collectors, particularly tropical forest dwellers and desert nomadic pastoralists. These distinctive cultures are eroding as a result of encroachment and resettlement of people from heavily populated areas. Their knowledge is particularly important in that they not only forage and collect, but manage forest vegetation in subtle, almost invisible, and usually ignored patterns of resource use. They have acquired a great deal of information about medicinals and many other plant and animal uses, and little Western scientific information exists on the sustainable use of these areas.
- Coastal fisherman, strand foragers, and small island villagers, such as the Orang Laut and Moken in Asia. These are the marine analogues of the forest foragers, their societies having evolved to take advantage of aquatic resources. They too are experiencing rapid resettlement, with consequent loss of their specialized knowledge. The often unique nature of coral reef and island ecosystems in the tropics makes this group of especial importance.
- Subsistence agriculturalists raising nonconventional staple crops and animals, as well as subsistence agriculturalists raising local cultivars and breeds of conventional crops and animals. There are still many groups—for example, the Senoi villages in Peninsular Malaysia, the Haya in the Usumbura mountains of Tanzania, the Aymara of the altiplano around Lake Titicaca in Peru and Bolivia, and many Amazon

groups as well as those in the Andean highlands (NRC, 1989)—that grow nonconventional crops, indigenous cereals, legumes and tubers on a small scale, which are better suited than exotic crops to local environmental conditions. Although the traditional resource use patterns of such groups are not changing as rapidly as some of those mentioned earlier, the economic pressure to produce only plant and animal varieties marketed internationally is rapidly eroding the genetic base of many important world food commodities. The diversity of genetic stocks, species composition, and management methods that characterizes and has long sustained traditional agricultural systems is being lost.

- Groups that have successfully adapted traditional technologies and resource use patterns in developing market opportunities, such as the ribereños of the Peruvian Amazon.

A Cultural Research Agenda

Social science research should improve our understanding of the relationship between biological diversity and local knowledge bases and of the activities that determine resource use, and should translate this understanding into policy and program tools. The central question that social sciences can help illuminate in the effort to conserve biodiversity is, How do local people affect the biological diversity of the ecosystems they inhabit? More specifically,

- How do local people use ecological resources, and why?
- What effects do these uses have on biological diversity?
- What changes in the conditions of local people promote patterns of use that deplete or conserve biological diversity?

Together, these questions provide a conceptual framework for more detailed investigations of the social and cultural aspects of biological diversity and its conservation.

Use of Resources—Social Concepts of Biodiversity

Research on the relationship among cultural patterns, economic bases, social activities, and the use of natural resources provides us with baseline data with which to interpret impacts, positive and negative, on the biological diversity of a given ecosystem. To understand more fully how and why local people use resources, researchers should do the following:

- Record the local knowledge system, including the species, communities, or ecosystems used; the quantity and type of products obtained;

the management systems employed; and reasons for choosing particular species.

- Determine how farmers, foragers, and other local men and women conceive of biodiversity, conservation, sustainability, and genetic erosion, and the historical or cultural basis of their views.
- Identify conservation practices used, the behavioral basis for those practices, and the role of social institutions (such as kin groups, religion, or belief systems) in the employment of these practices.
- Describe the theories, implicit or embedded, that people use to guide modification of ecosystems to achieve certain goals.
- Determine the scale dependence of local technologies, the degree to which they can be employed on a broader scale, and the modifications and cautions that must accompany such applications.
- Devise simple methods for assessing the genetic diversity of local crops and animals, and means by which this information can be used to add to local understanding of genetic variability.

Effects of Land or Resource Tenure and Uses on Biological Diversity

Research on the effect of traditional and nontraditional resource use on biological diversity builds on the baseline information outlined above to help us understand the relationship between social actions and ecological consequences. To understand more fully this relationship, researchers should:

- Determine how resource use patterns may have evolved prior to the introduction of recent external influences, and the degree to which this altered and formed the diversity characteristic of the locale;
- Examine the penetration of external farming technologies into local agricultural systems and assess the impact of these technologies on decisions about the conservation or loss of local cultivars and management tools;
- Determine how the degree of landscape fragmentation affects the organization and content of local knowledge; and
- Assess the degree to which local institutions and belief systems have or have not served as effective agents for conservation of biodiversity.

Impact of Changing Conditions

Research on the changing conditions of local people,* and the effect of these changes on resource use patterns, is necessary if we are to

*This research will be site specific and, for the most part, only locally relevant in its details; "local people" are thus not generic in this case.

understand more fully the basic causes and consequences of biodiversity loss. This research essentially involves the relationship between local actions and supralocal influences, and the impact of that changing relationship on the status of biodiversity. To understand more fully the dynamics of this relationship, researchers should:

- Identify mechanisms within local knowledge or decision-making systems that allow local peoples to react and adapt to exogenous factors;
- Characterize the relationship between degree of local control of resources and conservation of biodiversity, and how such relationships can further conservation goals (e.g., by encouraging the observance of reserve boundaries through social incentives);
- Investigate the structure and function of local organizations whose decisions affect biodiversity and determine how local organizations that support the conservation of biodiversity can be strengthened; and
- Describe the social factors that should be considered when establishing reserves in areas that are inhabited or adjacent to inhabited areas.

Social Valuation of Biodiversity at the National Level

At this level, the following steps must be taken:

- Evaluate the adequacy of the institutional infrastructure charged with managing natural resources and implementing conservation programs.
- Describe the status of communication between national institutions charged with implementation of conservation programs and local institutions involved in the use of biological resources.
- Identify how the objectives of agricultural and environmental officers in development agencies diverge, and the effect of this divergence on conservation programs.
- Identify local and national institutions that could be used to enhance the national status of local knowledge. Understand how local institutions deal with, or are altered by, the massive shifts in landscape use resulting from large development projects.
- Determine how national and international conservation priorities can be reconciled when they are at odds (Peru, for example, may want to concentrate on saving high-elevation crop germ plasm, whereas the World Wildlife Fund will give higher priority to the Peruvian Amazon).
- Determine how national goals of increasing foreign exchange through exploitation of the natural resource base can be reconciled with local goals of using resources in a sustainable manner.

- Determine why development agencies, national governments, urban elites, and educational systems devalue local knowledge, and identify measures that can be taken to reverse this situation.
- Describe the structural and philosophical changes that would allow development agencies to use local knowledge more effectively in development activities.
- Identify and characterize local and national institutions that currently record local knowledge and are most appropriate for recording and disseminating it.
- Determine the intellectual property rights issues associated with local knowledge and whether local or national institutions can ensure these rights.

References and Recommended Reading

Abramovitz, J.N. 1991. Investing in Biological Diversity: U.S. Research and Conservation Efforts in Developing Countries. Washington, D.C.: World Resources Institute.

Alcorn, J. and C. Hernàndez V. 1983. Plants of the Huastecan region of Mexico with analysis of their Huastec names. J. Mayan Ling. 4:11-118.

Allen, T.F.H., and T.B. Starr. 1982. Hierarchy: Perspectives for Ecological Complexity. Chicago: University of Chicago Press.

Amaranthus, M.P., and D.A. Perry. 1987. The effect of soil inoculation on ectomycorrhizal formation and the survival and growth of conifer seedlings on old, non-reforested clearcuts. Can. J. For. Res. 17:944-950.

Ashton, P. 1989. Funding Priorities for Research Towards Effective Sustainable Management of Biodiversity Resources in Tropical Asia. Report of a workshop sponsored by NSF and USAID held in Bangkok, Thailand, March 27–30, 1989. Unpublished report.

Berlin, B., D.E. Breedlove, and P. Raven. 1974. Principles of Tzeltal Plant Classification: An Introduction to the Botanical Ethnography of a Mayan-Speaking People of Highland Chiapas. New York: Academic Press.

Böhlke, J.E., S.H. Weizman, and N.A. Manezes. 1978. Estado atual da sistemática dos peixes de água doce da América do Sul. Acta Amazonica 8(4):657-677.

Bormann, F.H., and G.E. Likens. 1979. Pattern and Process in a Forested Ecosystem. New York: Springer-Verlag.

Braden, J.B., and C.D. Kolstad, eds. 1991. Measuring the Demand for Environmental Quality. New York: Elsevier.

Brush, S.B., H.J. Carney, and Z. Muamán. 1991. Dynamics of Andean potato agriculture. Economic Botany 35:70-88.

Burley, W.F. 1988. The tropical forest action plan: Recent progress and new initiatives. Pp. 403–408 in Biodiversity, E.O. Wilson and F.M. Peter, eds. Washington, D.C.: National Academy Press.

Cairns, Jr., J. 1988. Increasing diversity by restoring damaged ecosystems. Pp. 333–343 in Biodiversity, E.O. Wilson and F.M. Peter, eds. Washington, D.C.: National Academy Press.

Charney, J.G. 1975. Dynamics of deserts and drought in the Sahel. Quarterly Journal of the Royal Meteorological Society 101 (428):193-202.

de Chassey, F. 1978. La Mauritanie—1900–1975. Paris: Editions Anthropos.

Christonty, L. 1986. Shifting cultivation and tropical soils: Patterns, problems, and possible improvements. Pp. 226–240 in Traditional Agriculture in South East Asia: A Human Ecology Perspective, G.G. Martin, ed. Honolulu, Hawaii: East-West Center.

Conklin, H.C. 1980. Ethnographic Atlas of Ifugao. New Haven, Conn.: Yale University Press.

Cook, R.J. 1991. Challenges and rewards of sustainable agriculture research and education. Pp. 32–76 in Sustainable Agriculture Research and Education in the Field. Washington, D.C.: National Academy Press.

Costanza, R., ed. 1991. The Science and Management of Sustainability. New York: Columbia University Press.

Costanza, R., and C. Perrings. 1990. A flexible assurance bonding system for improved environmental management. Ecological Economics 2:57-76.

Cremoux, P. 1963. The importance of game-meat consumption in the diet of sedentary and nomadic peoples of the Senegal River Valley. Pp. 127–129 in Conservation of Nature and Natural Resources in Modern African States. IUCN Publications New Series, No. 1. Morges, Switzerland: IUCN.

Daly, H.E., and J.B., Cobb, Jr. 1990. For the Common Good: Redirecting the Economy Toward Community, the Environment, and a Sustainable Future. Boston: Beacon Press.

Demain, A.L., and N.A. Solomon. 1981. Industrial microbiology. Scientific American 245(3):66-75.

Ecological Society of America (ESA). 1991. The sustainable biosphere initiative: An ecological research agenda. Ecology 72(2):371-412.

Ehrlich, P.R., and E.O. Wilson. 1991. Biodiversity studies: Science and policy. Science 253:758-762.
Erwin, T.L. 1982. Tropical forests: Their richness in Coleoptera and other arthropod species. Coleopterists' Bulletin 36:74-75.
Erwin, T.L. 1983. Tropical forest canopies: The last biotic frontier. Bulletin of the Entomological Society of America 29(1):14-19.
Erwin, T.L. 1991. How many species are there? Revisited. Conservation Biology 5:330-333.
Franklin, J.F. 1989. Importance and justification of long-term studies in ecology. PP. 3–19 in Long-Term Studies in Ecology: Approaches and Alternatives, G.E. Likens, ed. New York: Springer-Verlag.
Franklin, J.F., C.S. Bledsoe, and J.T. Callahan. 1990. Contributions of long-term ecological research program: An expanded network of scientists, sites, and programs can provide crucial comparative analyses. Bioscience 40(7):509-523.
Gaston, A., and D. Dulieu. 1976. Pâturages du Kanem. Maisons-Alfort, France: Institut d'Élevage et de Médecine Vétérinaire des Pays Tropicaux.
Gaston, K.J. 1991. The magnitude of global insect species richness. Conservation Biology 5(3):283-296.
Gómez-Pompa, A. 1987. On Maya silviculture. Mexican Studies/Estudios Mexicanos 3(1):1-17.
Gómez-Pompa, A. 1988. Tropical deforestation and Maya silviculture: An ecological paradox. Tulane Studies in Zoology and Botany 26:19-37.
Grainger, A. 1984. Quantifying changes in forest cover in the humid tropics: Overcoming current limitations. Journal of World Forest Resource Management 1:3-63.
Grainger, A. 1988. Estimating areas of degraded tropical lands requiring replenishment of forest cover. International Tree Crops Journal 5:31-61.
Green, K.M. 1981. Digital processing of tropical forest habitat in Bangladesh and the development of low cost processing facility at the National Zoo, Smithsonian Institution. Proceedings of 15th International Symposium on Remote Sensing of Environment 3:1315-1325.
Green, K.M., J.F. Lynch, J. Sircar, and L.S.Z. Greenberg. 1987. Landsat remote sensing to assess habitat for migratory birds in the Yucatan peninsula, Mexico. Vida Silvestre Neotropical 1(2):27-38.
Guzman, H.M. 1991. Restoration of coral reefs in Pacific Costa Rica. Conservation Biology 5(2):189-195.
Harley, J.L., and S.E. Smith. 1983. Mycorrhizal Symbiosis. New York: Academic Press.
Harris, L.D. 1984. The Fragmented Forest: Island Biogeography Theory and the Preservation of Biotic Diversity. Chicago: University of Chicago Press.
Heady, H.F., and E.B. Heady. 1982. Range and Wildlife Management in the Tropics. London: Longman Press.
Houghton, J.T., G.J. Jenkins, and J.J. Ephraums, eds. 1990. Climate Change: The IPCC Scientific Assessment. Cambridge, England: Cambridge University Press.
Jablonski, D. 1991. Extinctions: A paleontological perspective. Science 253:754-757.
Janzen, D.H. 1988. Tropical dry forests: The most endangered major tropical ecosystem. Pp. 130–137 in Biodiversity, E.O. Wilson and F.M. Peter, eds. Washington, D.C.: National Academy Press.
Jenkins, Jr., R.E. 1988. Heritage Conservation Data Center. Paper presented to IUCN General Assembly, February 4, 1988.
Jordan III, W.R. 1988. Ecological restoration: Reflections of a half-century of experience at the University of Wisconsin-Madison Arboretum. Pp. 311-316 in Biodiversity, E.O. Wilson and F.M. Peter, eds. Washington, D.C.: National Academy Press.
Jordan III, W.R., M.E. Gilpin, and J.D. Aber. 1987. Restoration Ecology: A Synthetic Approach to Ecological Research. New York: Cambridge University Press.
Kneese, A.V. 1984. Measuring the Benefits of Clean Air and Water. Washington, D.C.: Resources for the Future.
Kneese, A.V., and B.T. Bower. 1968. Managing Water Quality: Economics, Technology, Institutions. Washington, D.C.: Resources for the Future.
Kristensen, R.M. 1983. Loricifera, a new phylum with Aschelminthes characters from the meiobenthos. Z. Zool. Syst. 21(3):163-180.
Likens, G.E., ed. 1989. Long-Term Studies in Ecology: Approaches and Alternatives. New York: Springer-Verlag.
Lorini, M.L., and V.G. Persson. 1990. New species of Leontopithecus lesson 1840 from Southern Brazil primates Callitrichidae. Bol. Mus. Nac. Rio. J. Zool. O. 338:1-14.
Luoma, J.R. 1991. A wealth of forest species is underfoot. New York Times, July 2, 1991. C1; C9.
Malhotra, K.C., S.B. Khomne, and M. Gadgil. 1983. Hunting strategies among three non-pastoral nomadic groups of Maharashtra. Man in India 63(1):23-39.
Martin, P.S. 1984. Prehistoric overkill: The global model. Pp. 354–403 in Quaternary Extinctions: A Prehistoric Revolution, P.S. Martin and R.G. Klein eds. Tucson, Ariz.: University of Arizona Press.

Mayr, E. 1982. The Growth of Biological Thought: Diversity, Evolution, and Inheritance. Cambridge, Mass.: Belknap Press.
McNeely, J.A. 1988. Economics and Biological Diversity: Developing and Using Economic Incentives to Conserve Biological Resources. Gland, Switzerland: IUCN.
McNeely, J.A. 1989. Priorities in conservation biology. Conservation Biology 3:416-417.
McNeely, J.A., K.R. Miller, W.V. Reid, R.A. Mittermeier, and T.B. Werner. 1990. Conserving the World's Biological Diversity. Gland, Switzerland and Washington, D.C.: IUCN, WRI, CI, WWF-US, and the World Bank.
Matson, M., and B. Holben. 1987. Satellite detection of tropical burning in Brazil. International Journal of Remote Sensing 8(3):509-516.
May, R.M. 1988. How many species are there on the Earth? Science 241:1441-1449.
Miller, R.M. 1985. Mycorrhizae. Restoration and Management Notes 3(1):14-20.
Mitchell, R.C., and R.T. Carson. 1989. Using Surveys to Value Public Goods: The Contingent Valuation Method. Washington, D.C.: Resources for the Future.
Morin, N.R., R.D. Whetsonte, D. Wilken, and K.L. Tomlinson. 1989. Floristics for the 21st Century: Proceedings of the Workshop 4–7 May, 1988, Alexandria, Va. Monographs in Systematic Botany from the Missouri Botanical Garden. Volume 28. Ann Arbor, Mich.: Braun-Brumfield.
Myers, N. 1988. Threatened biotas: Hotspots in tropical forests. Environmentalist 8(3):1-20.
National Academy of Sciences (NAS). 1980. Research Priorities in Tropical Biology. Washington, D.C.: National Academy of Sciences.
National Research Council (NRC). 1975. Underexploited Tropical Plants with Promising Economic Value. BOSTID Report 16. Washington, D.C.: National Research Council.
National Research Council. 1979. Tropical Legumes: Resources for the Future. BOSTID Report 25. Washington, D.C.: National Research Council.
National Research Council. 1982. Ecological Aspects of Development in the Humid Tropics. Washington, D.C.: National Academy Press.
National Research Council. 1983. Environmental Change in the West African Sahel. Washington, D.C.: National Academy Press.
National Research Council. 1986a. Ecological Knowledge and Environmental Problem-Solving: Concepts and Case Studies. Washington, D.C.: National Academy Press.
National Research Council. 1986b. Proceedings of the Conference on Common Property Resource Management. Washington, D.C.: National Academy Press.
National Research Council. 1989. Science and Technology Information Services and Systems in Africa. Washington, D.C.: National Academy Press.
National Research Council. 1990. The Improvement of Tropical and Subtropical Rangelands. Washington, D.C.: National Academy Press.
National Research Council. 1991a. Managing Global Genetic Resources: Forest Trees. Washington, D.C.: National Academy Press.
National Research Council. 1991b. Managing Global Genetic Resources: The U.S. National Plant Germplasm System. Washington, D.C.: National Academy Press.
National Research Council. 1991c. Policy Implications of Greenhouse Warming. National Research Council. Washington, D.C.: National Academy Press.
National Research Council. 1991d. Towards Sustainability: A Plan for Collaborative Research on Agriculture and Natural Resource Management. Washington, D.C.: National Academy Press.
National Science Board (NSB). 1989. Loss of Biological Diversity: A Global Crisis Requiring International Solutions. Report NSB-89-171. Washington: D.C.: National Science Foundation.
Nepstad, D.C., C. Uhl, and A.S. Serrao. 1991. Recuperation of a degraded Amazonian landscape: Forest management and agricultural restoration. Ambio 20(6):248-255.
Neuenschwander, P., W.N.O. Hammond, O. Ajuonu, A. Gado, N. Eschendu, A.H. Bokonon-Ganta, R. Allomasso, and I. Okon. 1990. Biological control of the cassava mealybug, Phenacoccus manihoti (Hom., Pseudococcidae) by Epidinocarsis lopezi (Hym., Encyrtidae) in West Africa, as influenced by climate and soil. Agriculture, Ecosystems, and Environment 32:39-55.
Norgaard, R. 1988. The biological control of cassava mealybug in Africa. American Journal of Agricultural Economics 70(2):366-371.
Norgaard, R. 1991. Sustainability as Intergenerational Equity: The Challenge to Economic Thought and Practice. World Bank Discussion Paper. Washington, D.C.: World Bank.
Norse, E.A. 1990. Ancient Forests of the Pacific Northwest. Washington, D.C.: Island Press.
Noss, R.F. 1990. Indicators for monitoring biodiversity: A hierarchical approach. Conservation Biology 4(4):355-364.
O'Neill, R.V., D.L. DeAngelis, J.B. Waide, and T.F.H. Allen. 1986. A Hierarchical Concept of Ecosystems. Princeton, N.J.: Princeton University Press.
Orians G.H., ed. 1991. The Preservation and Valuation of Biological Resources. Seattle, Wash.: University of Washington Press.
Padoch, C. 1986. Site selection among permanent-field farmers: An example in East Kalimantan,

Indonesia. Journal of Ethnobiology: 6(2)279-288.
Panayotou, T. 1989. Economics of Environmental Degradation Problems, Causes and Responses. Harvard Institute for International Development Report prepared for the U.S. Agency for International Development under CAER Task Order #3.
Panayotou, T. 1990. Counting the cost: Resource degradation in the developing world. Fletcher Forum of World Affairs 14(2):270-283.
Pearce, D.W., and R.K. Turner. 1990. Economics of Natural Resources and the Environment. Baltimore, Md.: Johns Hopkins Press.
Peters, C.M., A.H. Gentry, and R. Mendelsohn, 1989. Valuation of a tropical forest in Peruvian Amazonia. Nature 339:655-656.
Population Reference Bureau (PRB). 1989. World Population Data Sheet (computer diskette). Washington, D.C.: Population Reference Bureau.
Pyne, S.J. 1991. Burning Bush: A Fire History of Australia. New York: Henry Holt.
Randall, A. 1991a. Nonuse Benefits. In: Measuring the Demand for Environmental Quality, J.B. Braden and C.D. Kolstad, eds. New York: Elsevier.
Randall, A. 1991b. Thinking about the value of biodiversity. Unpublished manuscript.
Raven, P. 1988. Biological resources and global stability. Pp. 3–27 in Evolution and Coadaptation in Biotic Communities, S. Kawano, J.H. Connell, and T. Hidaka, eds. Tokyo: University of Tokyo Press.
Ray, G.C. 1988. Ecological diversity in coastal zones and oceans. Pp. 36–50 in Biodiversity, E.O. Wilson and F.M. Peter, eds. Washington, D.C.: National Academy Press.
Red Latinomericana de Botànica (RLB). 1991. Report of Activities, 1990. Santiago: RLB.
Reid, W.V., and K.R. Miller. 1989. Keeping Options Alive: The Scientific Base for Conserving Biodiversity. Washington, D.C.: World Resources Institute.
Repetto, R. 1988. The Forest for the Trees? Government Policies and the Misuse of Forest Resources. Washington, D.C.: World Resources Institute.
Repetto, R., and M. Gillis, eds. 1988. Public Policies and the Misuse of Forest Resources. Cambridge, England: Cambridge University Press.
Risser, P.G., and J.M. Melillo. 1991. LTER: An International Proposal. New York: John Wiley & Sons.
Roberts, L. 1991. Ranking the rain forest. Science 251:1559-1560.
Robinson, M.H. 1988. Are there alternatives to destruction? Pp. 355–360 in Biodiversity, E.O. Wilson and F.M. Peter, eds. Washington, D.C.: National Academy Press.
Sader, S.A., T.A. Stone, and A.T. Joyce. 1990. Remote sensing of tropical forests: Overview of research and applications using non-photographic sensors. Photogrammetric Engineering and Remote Sensing 56(10):1343-1351.
Salam, A. 1989. Notes on Science, Technology and Science Education in the Development of the South. Trieste, Italy: Third World Academy of Sciences.
Sale, J.B. 1981. The Importance and Values of Wild Plants and Animals in Africa. Gland, Switzerland: IUCN.
Salvat, B. 1987. Human Impacts on Coral Reefs: Facts and Recommendations. French Polynesia: Antennes Museum E.P.H.E.
Saunders, D.A., R.J. Hobbs, and C.R. Margules. 1991. Biological consequences of ecosystem fragmentation: A review. Conservation Biology 5(1):18-32.
Scott, J.M., B. Csuti, and F. Davis. 1991a. Gap analysis: An application of global information systems for wildlife species. Pp. 167–180 in Challenge in the Conservation of Bioresources: A Practitioner's Guide, O.J. Decker, M.E. Krasny, F.R. Groff, C.R. Smith, and D.W. Gross, eds. Boulder, Colo.: Westview.
Scott, J.M., B. Csuti, K. Smith, J.E. Estes, S. Caicco. 1991b. Gap analysis of species richness and vegetation cover: An integrated biodiversity strategy. Pp. 282–297 in Balancing on the Brink of Extinction, Kathryn A. Kohm, ed. Washington, D.C.: Island Press.
Solbrig, O., ed. 1991. From Genes to Ecosystems: A Research Agenda for Biodiversity. Cambridge, Mass.: IUBS, SCOPE, and Unesco.
Sondaar, P.Y. 1977. Insularity and its effect on mammal evolution. Pp. 671–707 in Major Patterns in Vertebrate Evolution, M.K. Hecht, R.C. Goody, and B.M. Hecht, eds. New York: Plenum.
Soulé, M.E. 1986. Conservation Biology: The Science of Scarcity and Diversity. Sunderland, Mass.: Sinauer Associates.
Soulé, M.E. 1991. Conservation: Tactics for a constant crisis. Science 253:744-750.
Soulé, M.E., and K.A. Kohm, eds. 1989. Research Priorities for Conservation Biology. Washington, D.C.: Island Press.
Soulé, M.E., and B.A. Wilcox, eds. 1980. Conservation Biology: An evolutionary-ecological perspective. Sunderland, Mass.: Sinauer Associates.
Stork, N.E. 1988. Insect Diversity: Facts, fiction, and speculation. Biological Journal of the Linnean Society. 35:321-337.
Tangley, L. 1990. Cataloguing Costa Rica's diversity. Bioscience 40:633-636.

Tatum, L.A. 1971. The southern corn leaf blight epidemic. Science 171:1113-1116.
Thomas, G.W. 1980. The Sahelian/Sudanian Zones of Africa: Profile of a Fragile Environment. Report to the Rockefeller Foundation. New York: Rockefeller Foundation.
Thorne-Miller, B., and J. Cantena. 1991. The Living Ocean: Understanding and Protecting Marine Biodiversity. Washington, D.C.: Island Press.
Uhl, C. 1988. Restoration of degraded lands in the Amazon basin. Pp. 326–332 in Biodiversity, E.O. Wilson and F.M. Peter, eds. Washington, D.C.: National Academy Press.
Uhl, C., D. Nepstad, R. Buschbach, K. Clark, B. Kauffman, and S. Subler. 1990. Studies of ecosystem response to natural and anthropogenic disturbances provide guidelines for designing sustainable land-use systems in Amazonia. Pp. 24–42 in Alternatives to Deforestation: Steps toward Sustainable Use of the Amazon Rain Forest, A.B. Anderson, ed. New York: Columbia University Press.
United Nations Educational Scientific, and Cultural Organization (Unesco). 1974. Task Force on Criteria and Guidelines for the Choice and Establishment of Biosphere Reserves. Final Report. MAB Report Series No. 22. Paris: Unesco.
U.S. Congress, Office of Technology Assessment (OTA). 1987. Technologies to Maintain Biological Diversity. Washington, D.C.: U.S. Government Printing Office.
U.S. Environmental Protection Agency (EPA). 1985. A Methodological Approach to an Economic Analysis of the Beneficial Outcomes of Water Quality. Washington, D.C.: U.S. Environmental Protection Agency.
Vibulsresth, S. 1986. Remote sensing activities in Thailand. Pp. 127–138 in Remote-Sensing Yearbook. London: Taylor & Francis.
Water Resources Council. 1980. Principles and Guidelines for Planning and Related Land Resources. Washington, D.C.: U.S. Government Printing Office.
Wilson, E.O. 1987. The arboreal ant fauna of Peruvian Amazon forests: A first assessment. Biotropica 2:245-251.
Wilson, E.O. 1988. The current state of biological diversity. Pp. 3–18 in Biodiversity, E.O. Wilson and F.M. Peter, eds. Washington, D.C.: National Academy Press.
Wilson, E.O. 1989. Threats to biodiversity. Scientific American 261(3):108.
Wilson, E.O., and F.M. Peter, eds. 1988. Biodiversity. Washington, D.C.: National Academy Press.
World Bank. 1989. Sub-Saharan Africa: From Crisis to Sustainable Growth. Washington, D.C.: World Bank.
World Commission on Environment and Development (WCED). 1987. Our Common Future. Oxford, England: Oxford University Press.
World Resources Institute (WRI). 1986. World Resources 1986. New York: Basic Books.
World Resources Institute. 1987. World Resources 1987. New York: Basic Books.
World Resources Institute. 1991. Compact for a New World. Washington, D.C.: World Resources Institute.

Board on Science and Technology for International Development
Publications and Information Services (FO-2060Z)
Office of International Affairs
National Research Council
2101 Constitution Avenue, N.W.
Washington, D.C. 20418 USA

How to Order BOSTID Reports

BOSTID manages programs with developing countries on behalf of the U.S. National Research Council. Reports published by BOSTID are sponsored in most instances by the U.S. Agency for International Development. They are intended for distribution to readers in developing countries who are affiliated with governmental, educational, or research institutions, and who have professional interest in the subject areas treated by the reports.

BOSTID books are available from selected international distributors. For more efficient and expedient service, please place your order with your local distributor. Requestors from areas not yet represented by a distributor should send their orders directly to BOSTID at the above address.

Energy

33. **Alcohol Fuels: Options for Developing Countries.** 1983, 128 pp. Examines the potential for the production and utilization of alcohol fuels in developing countries. Includes information on various tropical crops and their conversion to alcohols through both traditional and novel processes. ISBN 0-309-04160-0.

36. **Producer Gas: Another Fuel for Motor Transport.** 1983, 112 pp. During World War II Europe and Asia used wood, charcoal, and coal to fuel over a million gasoline and diesel vehicles. However, the technology has since been virtually forgotten. This report reviews producer gas and its modern potential. ISBN 0-309-04161-9.

56. **The Diffusion of Biomass Energy Technologies in Developing Countries.** 1984, 120 pp. Examines economic, cultural, and political factors that affect the introduction of biomass-based energy technologies in developing countries. It includes information on the opportunities for these technologies as well as conclusions and recommendations for their application. ISBN 0-309-04253-4.

Technology Options

14. **More Water for Arid Lands: Promising Technologies and Research Opportunities.** 1974, 153 pp. Outlines little-known but promising technologies to supply and conserve water in arid areas. ISBN 0-309-04151-1.

21. **Making Aquatic Weeds Useful: Some Perspectives for Developing Countries.** 1976, 175 pp. Describes ways to exploit aquatic weeds for grazing and by harvesting and processing for use as compost, animal feed, pulp, paper, and fuel. Also describes utilization for sewage and industrial wastewater. ISBN 0-309-04153-X.

34. **Priorities in Biotechnology Research for International Development: Proceedings of a Workshop.** 1982, 261 pp. Report of a workshop organized to examine opportunities for biotechnology research in six areas: 1) vaccines, 2) animal production, 3) monoclonal antibodies, 4) energy, 5) biological nitrogen fixation, and 6) plant cell and tissue culture. ISBN 0-309-04256-9.

61. **Fisheries Technologies for Developing Countries.** 1987, 167 pp. Identifies newer technologies in boat building, fishing gear and methods, coastal mariculture, artificial reefs and fish aggregating devices, and processing and preservation of the catch. The emphasis is on practices suitable for artisanal fisheries. ISBN 0-309-04260-7.

73. **Applications of Biotechnology to Traditional Fermented Foods**. 1992, 207 pp. Microbial fermentations have been used to produce or preserve foods and beverages for thousands of years. New techniques in biotechnology allow better understanding of these transformations so that safer, more nutritious products can be obtained. This report examines new developments in traditional fermented foods. ISBN 0-309-04685-8.

Plants

47. **Amaranth: Modern Prospects for an Ancient Crop.** 1983, 81 pp. Before the time of Cortez, grain amaranths were staple foods of the Aztec and Inca. Today this nutritious food has a bright future. The report discusses vegetable amaranths also. ISBN 0-309-04171-6.

53. **Jojoba: New Crop for Arid Lands.** 1985, 102 pp. In the last 10 years, the domestication of jojoba, a little-known North American desert shrub, has been all but completed. This report describes the plant and its promise to provide a unique vegetable oil and many likely industrial uses. ISBN 0-309-04251-8.

63. **Quality-Protein Maize.** 1988, 130 pp. Identifies the promise of a nutritious new form of the planet's third largest food crop. Includes chapters on the importance of maize, malnutrition and protein quality, experiences with quality-protein maize (QPM), QPM's potential uses in feed and food, nutritional qualities, genetics, research needs, and limitations. ISBN 0-309-04262-3.

64. **Triticale: A Promising Addition to the World's Cereal Grains.** 1988, 105 pp. Outlines the recent transformation of triticale, a hybrid between wheat and rye, into a food crop with much potential for many marginal lands. The report discusses triticale's history, nutritional quality, breeding, agronomy, food and feed uses, research needs, and limitations. ISBN 0-309-04263-1.

67. **Lost Crops of the Incas.** 1989, 415 pp. The Andes is one of the seven major centers of plant domestication but the world is largely unfamiliar with its native food crops. When the Conquistadores brought the potato to Europe, they ignored the other domesticated Andean crops—fruits, legumes, tubers, and grains that had been cultivated for centuries by the Incas. This book focuses on 30 of the "forgotten" Incan crops that show promise not only for the Andes but for warm-temperate, subtropical, and upland tropical regions in many parts of the world. ISBN 0-309-04264-X.

70. **Saline Agriculture: Salt-Tolerant Plants for Developing Countries.** 1989, 150 pp. The purpose of this report is to create greater awareness of salt-tolerant plants and the special needs they may fill in developing countries. Examples of the production of food, fodder, fuel, and other products are included. Salt-tolerant plants can use land and water unsuitable for conventional crops and can harness saline resources that are generally neglected or considered as impediments to, rather than opportunities for, development. ISBN 0-309-04266-6.

Innovations in Tropical Forestry

35. **Sowing Forests from the Air.** 1981, 64 pp. Describes experiences with establishing forests by sowing tree seed from aircraft. Suggests testing and development of the techniques for possible use where forest destruction now outpaces reforestation. ISBN 0-309-04257-7.

41. **Mangium and Other Fast-Growing Acacias for the Humid Tropics.** 1983, 63 pp. Highlights 10 acacia species that are native to the tropical rain forest of Australasia. That they could become valuable forestry

resources elsewhere is suggested by the exceptional performance of *Acacia mangium* in Malaysia. ISBN 0-309-04165-1.

42. **Calliandra: A Versatile Small Tree for the Humid Tropics.** 1983, 56 pp. This Latin American shrub is being widely planted by villagers and government agencies in Indonesia to provide firewood, prevent erosion, provide honey, and feed livestock. ISBN 0-309-04166-X.

43. **Casuarinas: Nitrogen-Fixing Trees for Adverse Sites.** 1983, 118 pp. These robust, nitrogen-fixing, Australasian trees could become valuable resources for planting on harsh eroding land to provide fuel and other products. Eighteen species for tropical lowlands and highlands, temperate zones, and semiarid regions are highlighted. ISBN 0-309-04167-8.

52. **Leucaena: Promising Forage and Tree Crop for the Tropics.** 1984 (2nd edition), 100 pp. Describes a multipurpose tree crop of potential value for much of the humid lowland tropics. Leucaena is one of the fastest growing and most useful trees for the tropics. ISBN 0-309-04250-X.

71. **Neem: A Tree for Solving Global Problems.** 1992, 149 pp. The neem tree offers great potential for agricultural, industrial, and commercial exploitation, and is potentially one of the most valuable of all arid-zone trees. It shows promise for pest control, reforestation, and improving human health. Safe and effective pesticides can be produced from seeds at the village level with simple technology. Neem can grow in arid and nutrient-deficient soils and is a fast-growing source of fuelwood. ISBN 0-309-04686-6.

Managing Tropical Animal Resources

32. **The Water Buffalo: New Prospects for an Underutilized Animal.** 1981, 188 pp. The water buffalo is performing notably well in recent trials in such unexpected places as the United States, Australia, and Brazil. Report discusses the animal's promise, particularly emphasizing its potential for use outside Asia. ISBN 0-309-04159-7.

44. **Butterfly Farming in Papua New Guinea.** 1983, 36 pp. Indigenous butterflies are being reared in Papua New Guinea villages in a formal government program that both provides a cash income in remote rural areas and contributes to the conservation of wildlife and tropical forests. ISBN 0-309-04168-6

45. **Crocodiles as a Resource for the Tropics.** 1983, 60 pp. In most parts of the tropics, crocodilian populations are being decimated, but programs in Papua New Guinea and a few other countries demonstrate that, with care, the animals can be raised for profit while protecting the wild populations. ISBN 0-309-04169-4.

46. **Little-Known Asian Animals with a Promising Economic Future.** 1983, 133 pp. Describes banteng, madura, mithan, yak, kouprey, babirusa, javan warty pig, and other obscure but possibly globally useful wild and domesticated animals that are indigenous to Asia. ISBN 0-309-04170-8.

68. **Microlivestock: Little-Known Small Animals with a Promising Economic Future.** 1990, 449 pp. Discusses the promise of small breeds and species of livestock for Third World villages. Identifies more than 40 species, including miniature breeds of cattle, sheep, goats, and pigs; eight types of poultry; rabbits; guinea pigs and other rodents; dwarf deer and antelope; iguanas; and bees. ISBN 0-309-04265-8.

Health

49. **Opportunities for the Control of Dracunculiasis.** 1983, 65pp. Dracunculiasis is a parasitic disease that temporarily disables many people in remote, rural areas in Africa, India, and the Middle East. Contains the findings and recommendations of distinguished scientists who were brought together to discuss dracunculiasis as an international health problem. ISBN 0-309-04172-4.

55. **Manpower Needs and Career Opportunities in the Field Aspects of Vector Biology.** 1983, 53 pp. Recommends ways to develop and train the manpower necessary to ensure that experts will be available in the future to understand the complex ecological relationships of vectors with human hosts and pathogens that cause such diseases as malaria, dengue fever, filariasis, and schistosomiasis. ISBN 0-309-04252-6.

60. **U.S. Capacity to Address Tropical Infectious Diseases.** 1987, 225 pp. Addresses U.S. manpower and institutional capabilities in both the public and private sectors to address tropical infectious disease problems. ISBN 0-309-04259-3.

Resource Management

50. **Environmental Change in the West African Sahel.** 1984, 96 pp. Identifies measures to help restore critical ecological processes and thereby increase sustainable production in dryland farming, irrigated agriculture, forestry and fuelwood, and animal husbandry. Provides baseline information for the formulation of environmentally sound projects. ISBN 0-309-04173-2.

51. **Agroforestry in the West African Sahel.** 1984, 86 pp. Provides development planners with information regarding traditional agroforestry systems—their relevance to the modern Sahel, their design, social and institutional considerations, problems encountered in the practice of agroforestry, and criteria for the selection of appropriate plant species to be used. ISBN 0-309-04174-0.

72. **Conserving Biodiversity:** A Research Agenda for Development Agencies. 1992. 127 pp. Reviews the threat of loss of biodiversity and its context within the development process and suggests an agenda for development agencies. ISBN 0-309-04683-1.

Forthcoming Books from BOSTID

Vetiver Grass: A Promising Plant for Controlling Soil Erosion. (1992) This study will evaluate the potential of vetiver, a little-known grass that seems to offer a practical solution for controlling soil loss. Hedges of this deeply rooted grass catch and hold back sediments. The stiff foliage acts as a filter that also slows runoff and keeps moisture on site, allowing crops to thrive when neighboring ones are desiccated. In numerous tropical locations, vetiver hedges have restrained erodible soils for decades and the grass—which is pantropical—has shown little evidence of weediness.

BOSTID Publication Distributors

United States:

Agribookstore
1611 N. Kent Street
Arlington, VA 22209

agAccess
PO Box 2008
Davis, CA 95617

Europe:

I.T. Publications
103-105 Southhampton Row
London WC1B 4HH
Great Britain

S. TOECHE-MITTLER
TRIOPS Department
Hindenburgstr. 33
6100 Darmstadt
Germany

T.O.O.L. Publications
Sarphatistraat 650
1018 AV Amsterdam
Netherlands

Asia:

Asian Institute of Technology
Library & Regional
Documentation Center
PO Box 2754
Bangkok 10501
Thailand

National Bookstore
Sales Manager
PO Box 1934
Manila
Philippines

University of Malaya Coop. Bookshop Ltd.
Universiti of Malaya
Main Library Building
59200 Kuala Lumpur
Malaysia

Researchco Periodicals
1865 Street No. 139
Tri Nagar
Delhi 110 035
India

China Natl Publications
Import & Export Corp.
PO Box 88F
Beijing
China

South America:

Enlace Ltda.
Carrera 6a. No. 51-21
Bogota, D.E.
Colombia

Africa:

TAECON
c/o Agricultural Engineering Dept
P.O. Box 170 U S T
Kumasi
Ghana

Australasia:

Tree Crops Centre
P.O. Box 27
Subiaco, WA 6008
Australia

For More Information

To receive more information about BOSTID reports and programs, please fill in the attached coupon and mail it to:

Board on Science and Technology for International Development
Publications and Information Services (FO-2060Z)
Office of International Affairs
National Research Council
2101 Constitution Avenue, N.W.
Washington, D.C. 20418 USA

Your comments about the value of these reports are also welcome.

- -

Name ____
Title ____
Institution ____________________________________
Street Address ________________________________
____
City ____
Country ____________________ Postal Code ______________

72

- -

Name ____
Title ____
Institution ____________________________________
Street Address ________________________________
____
City ____
Country ____________________ Postal Code ______________

72